AI AND THE END OF HUMANITY

AI AND THE END OF HUMANITY

*What Darwin Can Teach Us About
the Universe and Our Future*

JAMES B. MILES

Copyright © 2023 James B. Miles

The moral right of the author has been asserted.

Apart from any fair dealing for the purposes of research or private study, or criticism or review, as permitted under the Copyright, Designs and Patents Act 1988, this publication may only be reproduced, stored or transmitted, in any form or by any means, with the prior permission in writing of the publishers, or in the case of reprographic reproduction in accordance with the terms of licences issued by the Copyright Licensing Agency. Enquiries concerning reproduction outside those terms should be sent to the publishers.

Matador
Unit E2 Airfield Business Park,
Harrison Road, Market Harborough,
Leicestershire. LE16 7UL
Tel: 0116 2792299
Email: books@troubador.co.uk
Web: www.troubador.co.uk/matador
Twitter: @matadorbooks

ISBN 978 1803135 717

British Library Cataloguing in Publication Data.
A catalogue record for this book is available from the British Library.

Printed and bound by CPI Group (UK) Ltd, Croydon, CR0 4YY
Typeset in 11pt Adobe Garamond Pro by Troubador Publishing Ltd, Leicester, UK

Matador is an imprint of Troubador Publishing Ltd

CONTENTS

Acknowledgements		*ix*
1	Cosmic Dinosaurs, and the Eighth Transition	1
2	Darwin "Lets Us Down". Really?	29
3	E.T. Type I – A Single Inheritance Mechanism	53
4	E.T. Type II – A Dual Inheritance Mechanism	94
5	Reason, and the Race Of Devils Problem	117
6	Transhumanism, Plus The Existential Threat of AI	141
7	Type II Nature and Human Nature	172
8	The Three Flavours of Type II Existence	193
9	Stellar Irony, and our Importance to the Galactic Future	219
Bibliography		*235*
Index		*255*

ACKNOWLEDGEMENTS

My everlasting debt to George Williams runs throughout this book, but others are owed a more specific mention. It was Tom Fish, science reporter at the *Express*, who first got me thinking about writing this book. Tom was producing an article about the most recent Mars lander, and had asked to speak to me on the consequences of Darwin's understanding for extraterrestrial life. It was then my brother, Dr Chris Miles, who challenged me to think more widely, and to both consider the implications for artificial intelligence and to give some voice to the two alternative Darwinian traditions.

Having checked back, George and I did correspond briefly – twenty years ago now – regarding extending his updating of Darwin's work to both extraterrestrial intelligence and artificial intelligence, but I never really followed up on the analysis, preferring to concentrate on the implications for terrestrial and biological intelligence. Hence this book might not exist were it not for my being inspired by Tom and Chris. Furthermore, it was Chris' admonishment as regards reviewing all three extant

Darwinian traditions that opened up the realisation that each provided the same conclusion on cosmic dinosaurs, and the same conclusion on self-aware artificial intelligence. That really was something I had not expected to uncover.

Thanks to everyone at Matador, and to Ben Cameron at Cameron Publicity & Marketing. Thanks to Rod Mackenzie, Dr Yorick Rahman, and my sister-in-law Şebnem Zorlu-Miles for their advice. And my thanks to Jack Bream, one of my two wonderful nephews, for volunteering to handle the social media side, something I had, probably foolishly, looked upon with horror and decided simply to ignore.

1

COSMIC DINOSAURS, AND THE EIGHTH TRANSITION

> "... then that may suggest that complex life – that intelligent life – is extremely rare indeed in the universe. Maybe there is a profound bottleneck in the evolution of complex life in the Milky Way. And perhaps this is why we continue to bear the Great Silence."
>
> – **Brian Cox**, astrophysicist and broadcaster (2021)

In 1950 the nuclear physicist Enrico Fermi famously asked why – given the incalculably vast number of planetary systems and the apparently modest requirements for life – we have not yet been contacted by other intelligent life. Why "the Great Silence"? From the early nineteenth century onwards scientists had been at the forefront of speculation about contact with extraterrestrial intelligence. It was Carl Friedrich Gauss, sometimes described as the greatest mathematician since antiquity, who is often credited with the 1820 suggestion that intelligent life on the Moon or Mars could be signalled by building in the Siberian tundra a gigantic 10-mile to a side squares-and-triangle proof of Pythagoras's theorem (Raulin-Cerceau, 2010). Gauss' invention of the heliotrope would then inspire others to try to transmit messages into space using either angled or focused mirrors. In the early 1850s Thomas Huxley, the zoologist later to become celebrated as Darwin's bulldog, would himself be drawn into the "hot controversy" of intelligent aliens (Desmond 1997, p.204). And right through the nineteenth century and into the first decades of the twentieth century the giants of the science community continued to weigh in, particularly after the creation of radio, with the great electrical engineer Nikola Tesla pointing to disturbances in his electrical sensors, before spending his final years trying to answer back to the cosmos. A few years after Tesla claimed to have received interplanetary communication, Marconi made a similar declaration ("the messages have been distinct but unintelligible"), with reports that Thomas Edison thought Marconi offered "good grounds for the theory that inhabitants of other planets are trying to signal to us. ... Either they are our intellectual equals or our superiors" (ISSN, 1920). Even Albert Einstein, interviewed in January 1921 on the "mystic wireless", stated that "there is every reason to believe that Mars and other planets are inhabited", that we may assume intelligent creatures do exist "elsewhere in the universe", but that we should be looking out for their light ray communications, rather than expecting wireless messages from them (Einstein, 1921).

However, by the second quarter of the twentieth century it would become clear that there were no signals from the Moon, Mars, or Venus, indeed none detected even from outside our solar system, and what has become known as the Great Silence has continued to perplex scientists to this day. Many believe, quite reasonably, that given that the building blocks of the very simplest life appear to be so common, there must be a significant bottleneck – sometimes touched on as "the Great Filter" – sitting somewhere later on in the evolutionary steps to intelligent life. In both his

2014 and his 2021 television series for the BBC, the astrophysicist Brian Cox tries to explain: "So that means that most scientists, I think, suspect that we will find simple life somewhere out there. ... But we must be careful, because the story of life on this planet shows that the transition from single-celled life to complex life may not have been inevitable". If that's the case, continues Cox, then that may suggest that complex life, that intelligent life, is extremely rare indeed in the universe.

But note that Cox here recognises no effective distinction, no further transition, between what is termed *complex multicellular* life and *intelligent* life ("that complex life – that intelligent life – is extremely rare"). Yet Darwin did recognise a distinction, a great discontinuity, a vast transition, between complex multicellular life and intelligent life, and pursuing Darwin's thinking to its end-point leads to some profoundly different expectations. For Darwin, the galaxy is very probably teeming with plant and animal life. Please let me say that again. According to Darwin's own understanding of evolution, it is not that the galaxy just might contain other complex multicellular life, or even that the galaxy may well contain other complex life, it is that the galaxy is very probably teeming with complex multicellular life.

For Cox, as we have seen no evidence of cosmic intelligence it must be rare, and cosmic animal life will therefore be almost as rare, as it sits on the same continuity, within the same transition, an unbroken spectrum from very low animal intelligence to very high animal intelligence. For Darwin, cosmic animal life can be very common indeed, but because of a logical discontinuity, an evolutionary firebreak, if you like, it is cosmic intelligence that will always be rare. For Darwin – who will in this book be presenting us with a profoundly different explanation of the transition to intelligence than is currently appreciated – while millions upon millions of planets in the Milky Way will be crowded with plants and animals in a vast variety of different forms, both small and large, including giant dinosaur-like creatures, contemplative life can only ever come in two possible and rare patterns. Darwin's first form of non-terrestrial intelligent life will be ferociously indifferent, without meaning or purpose, while his second form of intelligent life will be thoughtful and less murderous although never fully rational. Plus there is, at least for Darwin, the cosmic evolutionary irony, the vast celestial joke, that humankind really is the best and the smartest the universe can ever get to naturally. Yet the very same evolutionary principles then predict that any pure machine intelligence will be driven by the need to eradicate us; indeed that any form of self-aware artificial intelligence – being necessarily based on Darwin's first form of intelligent life – will become locked into an existential struggle with humankind. That for Darwin we may now face the end of humanity within perhaps as little as a few decades, and unless we can take certain steps immediately.

THE MAJOR TRANSITIONS IN EVOLUTION

The "selfish gene" revolution, the biologists' modern explanation of evolution that sees natural selection as predominantly operating at the level of the smallest unit, the gene, rather than at Darwin's hypothesised level of the individual, is generally held to have started in 1966 with the publication of the American biologist George C. Williams' *Adaptation and Natural Selection*. Gene-selectionism, or as Williams put it "the formally disciplined use of the theory of genic selection for problems of adaptation" (1966, p.270), sees apparent individual selection reinterpreted as not what is good for an individual but as what is good for its genes. Since the fate of an individual and the fate of its genes are very closely – but not perfectly – linked, Darwin's individual selection is often for practical purposes synonymous with gene selection.

The second biologist generally recognised as the father of gene-selectionism is the late great English biologist John Maynard Smith. For their co-development of gene-centred evolutionary biology Williams and Maynard Smith were together awarded the Crafoord Prize in 1999, the biologists' equivalent of the Nobel, and also given out by the Royal Swedish Academy of Sciences.[1] And like both Darwin and Williams, Maynard Smith, who would separately go on to become the father of evolutionary game theory, also realised that the galaxy may well be packed with complex life, and that intelligence, not complexity, was the likely bottleneck in evolution, and was the answer to the Great Silence.

One of the last books Maynard Smith wrote before his death was his influential 1995 *The Major Transitions in Evolution*, with the Hungarian evolutionary theorist Eörs Szathmáry. Earlier in his vastly prolific career Maynard Smith had made significant mathematical contributions to at least one of the major transitions, the evolution of sex, and this book was a collaboration recognising the major advances that can happen when evolution suddenly discovers a profoundly new pathway. Maynard Smith and Szathmáry recognised eight major transitions across time, and which fundamentally involved changes in the way information is stored and transmitted between generations. The book was also an account of the evolution of complexity, and the idea that new coding methods have made possible more complex organisms, and which begins to touch on Darwin's insight of why multicellular life may be so common in the galaxy, yet deliberative life so rare.

---

[1] Although George C. Williams was my friend and mentor for more than a decade prior to his death, and he even wrote the foreword to an earlier book of mine (see Miles 2003), I cannot claim to have known Maynard Smith at all. We did correspond briefly in 1997, which was enough for me to confirm that he and George were, intellectually, two peas in a pod, and that he was very much continuing to follow Darwin's own line of logic and reasoning.

Major transition	Darwin / Genic selection	Kelvin / Sociobiological
1. Compartmentalised molecules	Accepts	Accepts
2. Transition to chromosomes	Accepts	Accepts
3. RNA world to DNA	Accepts	Accepts
4. Prokaryotes to eukaryotes	Accepts	Accepts
5. Evolution of sex	Accepts	Accepts
6. Evolution of multicellularity	Accepts	Accepts
7. Eusocial colonies	Accepts	Accepts
8. Dual inheritance system	Accepts – and bottleneck	Rejects – but bottleneck

Some of the above transitions are too nuanced, too wide ranging, and still too debated, to describe properly here in a work of pop science. Maynard Smith and Szathmáry did rewrite their 1995 volume, which had been aimed at professional biologists, in 1999 and as a somewhat simpler book for a more popular readership, *The Origins of Life*, but just consider, for example, the extraordinary fourth transition from prokaryote to eukaryote, and which took place perhaps two billion years ago. The authors remind us that in the early 1970s the biologist Lynn Margulis finally convinced the scientific community of the symbiotic origin of mitochondria and chloroplasts. Endosymbiotic theory says that organelles of eukaryotic cells including mitochondria and plastids are originally the chance cooperative alliance of free-living primitive prokaryotic cells. In other words, one simple free-living bacteria-like cell once ingested another simple free-living bacteria-like cell, but instead of the attacker absorbing the victim there was no digestion ("a kind of 'cellular indigestion'", 1999, p.61), leading to a more complicated symbiotic alliance of one within the other. This serendipitous transition from prokaryote to eukaryote later allowed complex multicellular life to begin. Serendipitous, as evolution has no foresight – "a transition may have opened up new possibilities for future evolution, but that is not why it happened" (p.25). Or consider the authors' fifth transition with the evolution of sexual reproduction from asexual reproduction, and which at first seems so wasteful and individually inefficient, but immensely speeded up the process of adaptation. They write: "Sexually produced offspring are all different, whereas parthenogenetically produced offspring are usually identical genetically. As the American George Williams pointed out, a parthenogenetic female is like a man who buys 100 tickets in a raffle, and finds that they all have the same number. It would be better, like a sexual female, to buy only 50 tickets, all with different numbers" (pp.84–5). Or there is their seventh transition, the evolution of eusocial colonies, where new directions in reproduction, different coefficients of relatedness, and morphological delineation, combined to allow both the coexistence of vastly greater group sizes and the presence of non-reproductive castes.

As Szathmáry and Maynard Smith put it in their review article in *Nature*, "There is no theoretical reason to expect evolutionary lineages to increase in complexity with time. ... Nevertheless, eukaryotic cells are more complex than prokaryotic ones, animals and plants are more complex than protists, and so on. This increase in complexity may have been achieved as a result of a series of major evolutionary transitions" (1995). The authors also noted that fully six of their eight transitions probably happened (and needed to happen) just once in a single lineage, while transition six from single-celled protists to multicellular life happened a few times, and transition seven, colonial animals with sterile castes, evolved many times. The authors also identified a number of properties common to the transitions; including that entities that were capable of independent replication before the transition could afterwards replicate only as part of the larger whole; that it is generally difficult, though not always impossible, to reverse the transitions once they had happened ("once sex had arisen ... sex is hard to abandon", 1999, p.25; but "irreversibility is not absolute", 1995, p.9); that the lower level parts will still sometimes seek to disrupt the workings of the larger organism; and that new ways of transmitting information have arisen over time (such as encoded protein synthesis, and epigenesis).

The transition table above contrasts the "Darwin / Genic selection" interpretation with the "Kelvin / Sociobiological" interpretation. The latter is named for both Lord Kelvin, the brilliant but arrogant nineteenth-century British physicist who was so hostile to Darwinian gradualism, and the more recent American tradition termed human sociobiology. We will separately have to consider the implications for intelligent life beyond our planet – and machine intelligence – assuming the sociobiological tradition to instead be correct, but it is worth noting that sociobiology, which refuses to recognise Darwin's dual inheritance mechanism as the eighth transition, could perhaps only have come out of late twentieth-century America. Sociobiology today prefers to be known as evolutionary psychology, or "EP", but Maynard Smith continued to refer to EP as "son of sociobiology" until he died, and because EP inherits the core conviction, indeed the seeming core biological mistake, of human sociobiology. This is the belief that Darwin got it wrong on human evolution, and that for sociobiologists human genetic evolution broke – *and uniquely broke* – the billion-year mould of the rest of nature. Although prefigured by theorists in mid to late nineteenth-century Britain including Kelvin, the palaeontologist Sir Richard Owen, and the father of behavioural genetics Sir Francis Galton, sociobiology emerged (or re-emerged) at Harvard University in the early 1970s. Throughout the book we will try to stick to the single name of sociobiology, rather than updating to "EP", not only because EP rests on the same core judgement, but because human sociobiology is ultimately the ur-language, the root language, in which so much of the last 150 years of human biological self-importance seems to be written.

EP, behavioural genetics, and race science are all sub-disciplines within the social (or "soft") science of psychology, so only human sociobiology, at least initially a biological discipline, can look to provide the "hard science" evolutionary skeleton from which they must all hang if they hope to make any kind of logical sense. Because as Theodosius Dobzhansky, shaper of the genetic language of variation that emerged with the early twentieth-century synthesis of Darwin and the work of Gregor Mendel, once put it: "Nothing in biology makes sense except in the light of evolution" (Dobzhansky, 1973). In a very real way it is Harvard University that becomes the chief architect to modern race science. From Galton's racial science, through Owen and Kelvin's attempts to both reject gradualism and bring direction into human evolution, and on to the Harvard psychologist Steven Pinker's more recent efforts to privilege the human animal, the linking theme is a wholesale rejection of Darwin's insight that humans evolved under *exactly the same* gradualist rules of natural selection as all other life on Earth.

Sociobiology fundamentally seems to want to lift humankind out of nature, or at least outside of the previous billion years of nature. Modern sociobiology started with a 1971 essay by a young Harvard postgraduate called Robert Trivers, when he had a paper published in the American *Quarterly Review of Biology*. That essay did not specifically claim that Darwin had got it wrong on the human animal, but it hypothesised without any evidence ("no direct evidence ... nor its genetic basis", Trivers 1971, p.48) a wholly new direction in Darwinian evolution. Trivers' article might still have come to nothing, were it not that a few years later a Harvard professor, Edward O. Wilson, would reanimate, and formally rename, this idea as his new discipline of human sociobiology. "Human decency is animal", Wilson claimed in 1975 (Wilson, 1975a); and then that "morality evolved as instinct" (1978, p.5). The major problem, though, is that morality had never evolved at the genetic level in the hundreds of millions of years of animal evolution, never previously evolved across all the billions of species that have lived on this planet.

Morality had not evolved over those hundreds of millions of years, hadn't evolved across those billions of species, because it seemingly cannot evolve at the biological level, which becomes Darwin's key insight once we turn attention to both extraterrestrial and artificial intelligence. As we shall see, morality cannot evolve when you have only one inheritance mechanism. Darwin realised this, and Williams and Maynard Smith later demonstrated it mathematically. Indeed, all the fathers of modern gene– and individual-level selection say the same thing. Williams and Maynard Smith may have won the Crafoord Prize for their work, but their work was in turn mathematically buttressed by the analysis of both Oxford's Bill Hamilton, the creator of "inclusive fitness" theory, and the former Manhattan Project scientist George Price, who would go on to develop evolutionary game theory with Maynard Smith, and would ally with Hamilton to re-derive the latter's work, now known as the Price equation (Price, 1970; Maynard Smith and Price, 1973). "Natural selection … implies concurrently a complete disregard for any values, either of individuals or of groups, which do not serve competitive breeding. This being so, the animal in our nature cannot be regarded as a fit custodian for the values of civilized man", wrote Hamilton (1971, p.83). And for Price human virtue could only be, as Price's biographer Oren Harman puts it, where a second inheritance mechanism "beat out their own nature" (Judah, 2013).

In this analysis of non-terrestrial life, this analysis of

extraterrestrial intelligence, and the subsequent analysis of non-natural, hence artificial, intelligence, it is somewhat important to recognise early on that sociobiology is ultimately throwing away the mechanisms that Darwin developed, substituting instead evolutionary mechanisms never before witnessed, substituting mechanisms on the basis of "no direct evidence", and substituting mechanisms that seemingly cannot possibly work mathematically (as the late Ed Wilson himself would grudgingly admit after 2007). This does not necessarily make human sociobiology wrong, of course, and in order to get the full evolutionary picture we shall also have to analyse the implications for extraterrestrial intelligence and artificial intelligence when assuming Darwin was the one who got it wrong, and the sociobiologists are the ones who got it right. But sociobiology undeniably seeks to lift humankind away from the orthodox pattern of nature, which is why in the modern academic world it perhaps could only have re-emerged at an American university, and through an American academic journal. Having already turned their backs on earlier "exceptionalist" claims that humanity evolved under unique rules, claims by the likes of Owen and Kelvin, late twentieth-century British and European academia was generally more cautious about considering any theory that privileges the human animal, particularly where it was being presented with no direct evidence, "nor its genetic basis". Extraordinary claims require extraordinary evidence, and if you wish to claim that Darwin misunderstood his own theory, if you wish to claim that humans broke from the orthodox rules of natural selection, you have to provide more than just speculation. Darwin "used to look upon it as a great weakness if one allowed wish to influence belief" (Desmond & Moore 1991, p.608).

As with Lord Kelvin one and a half centuries ago, readers might not like the notion that at least in the first analysis we appear to have evolved under exactly the same rules as all other life on Earth, but if you follow Darwin's own reasoning there is, and mathematically and logically there has to be, an *eighth* major transition in evolution. In fact, though, even the sociobiologists will end up accepting that there has to be a very unusual eighth transition in play, with all the extraordinary new implications this will hold for life off this planet, and intelligence anywhere, natural or mechanical.

THE PRINCIPLE OF PARSIMONY

Brian Cox continues his earlier explanation: "Everything we would call a complex living thing today shares the same basic structure – it's built out of cells called eukaryotic cells. ... And they're extremely different to the simple cells". So how did those cells come to exist? Cox asks. One popular theory "is that it was two simple cells merging together that formed what we'd recognise today as the complex cells in your body". Cox is of course referring to Lynn Margulis and others' work above, to endosymbiotic theory. The idea that one organism can get inside another and doesn't kill it, that they both survive to "produce something that's actually capable of doing magnificent things, things that are far more complicated and wonderful than the two simple building blocks can manage on their own seems to be … it seems to be unlikely, a 'fateful encounter'" (Cox, 2021).

1. Compartmentalised molecules	
2. Transition to chromosomes	
3. RNA world to DNA	
4. Prokaryotes to eukaryotes	
5. Evolution of sex	
6. Evolution of multicellularity	
7. Eusocial colonies	
8. Dual inheritance system	

Cox is understandably trying to back-solve for the Great Silence. If the building blocks of simple life are so abundant, but intelligent life so rare, there must be a bottleneck somewhere. Cox places the bottleneck at transition four, prokaryote to eukaryote, as, like the sociobiologists, he fails to recognise the eighth transition, at least as Darwin understood it. Because as we have already seen, Cox conflates, he adds together transition six, complex multicellular life, and transition eight, the dual inheritance mechanism that Darwin saw as the key to intelligent life ("that complex life – that intelligent life – is extremely rare").

Is Cox so wrong to do this? Are the sociobiologists so wrong in refusing to recognise a dual inheritance mechanism as the critical eighth transition, and irrespective of the fact that Darwin recognised such a transition? Well, firstly let us mention there is no firm evidence that transition four is a bottleneck, and Maynard Smith and Szathmáry take issue with this idea: "If we are right, the changes were forced on the early eukaryotes because of the loss of the rigid outer cell wall of prokaryotes"; "having lost its 'external skeleton', however, the cell was forced to do several things to compensate" (1999, pp.25, 62). How likely was the loss of that rigid outer cell wall, which may have triggered everything that was to follow? Conceivably quite likely. "A possible, though speculative, suggestion is that some bacterium evolved, as a competitive device, an antibiotic that blocked the synthesis of the cell wall of another bacterium: that is how some modern antibiotics (for example, penicillin) kill bacteria", write Maynard Smith and Szathmáry (p.63). Wall-less bacteria are extremely fragile, they tell us, and such losses may have occurred several times in evolution. "The most common outcome would have been the extinction of such lineages. ... But two lineages, related to one another, found novel remedies." Furthermore, eukaryotes generally require oxygen, so would have been at a significant disadvantage until photosynthesis permitted an oxygen-rich planet, what is called the Great

Oxidation Event, which only came about perhaps a billion and a half years after the first prokaryote.

Secondly, the eighth transition as Darwin understood it truly seems to be a bottleneck. All the other seven transitions either spread across the full evolutionary landscape, or at least spread fast across a number of evolutionary niches. Language, intelligence, and the emergence of a second inheritance system have been restricted to just one species of animal that has ever arisen on this planet, out of billions of species. And what is happening to that one species beyond transition six is very weird indeed, one of the weirdest occurrences in nature, and therefore almost by definition presents a potential evolutionary chokepoint to intelligence. Because now note something fascinating, and whether or not you want to accept Darwin's explanation for an eighth transition. Sociobiology is still highlighting the existence of an evolutionary bottleneck after transition six. Why? Because remember what sociobiology is based upon; the claim that when it came to our species nature tore up the rulebook and created evolutionary processes never even theoretically imagined before we came along. That switch purportedly reversed, or at the very least substantially changed, the rules of genic selection. Sociobiology is still invoking something happening that appears to be probabilistically vastly unlikely in the roll call of life. *Sociobiology is still giving us an exceedingly rare eighth transition*, albeit not a second inheritance mechanism, not language operating on a large and psychologically susceptible brain, but instead an almost complete rewriting of the rules of natural selection, and a process argued to be happening in just one species out of billions. Darwin and the sociobiologists may be saying different things, but they are both converging on something astonishing happening to just one species of complex life after transition six. Both Darwin and the sociobiologists are presenting us with the same evolutionary chokepoint. Both Darwin and the sociobiologists are ultimately offering us cosmic dinosaurs; both are still giving us dinosaurs everywhere.

Some readers might now want to ask: what if Darwin is getting it wrong and the sociobiologists are getting it wrong? Is there a possible third way that doesn't invoke either the power of a second inheritance mechanism or wholly new principles of gene-centred natural selection? As we shall also investigate and follow to its off-planet and AI end-points, there is another (also somewhat Harvard) theory that suggests a third option called genetic group selection, but please understand that this option likewise needs something unique to be happening to in effect just one species beyond transition six. *So the group-selectionists are also giving us an exceedingly rare eighth transition*, albeit not language operating on a large and psychologically susceptible brain, but instead the emergence of a new level for natural selection to operate at, and a process also effectively argued to be happening in just one species out of billions. To quote two of the main theorists within this third school, Elliott Sober and David Sloan Wilson, human groups have, like social insect colonies, "been interpreted as superorganisms for centuries", and biologists need some explanation for "why humans are ultrasocial" (1998, p.158). For Sober and D.S. Wilson, at least when it comes to group sizes, humans are behaving as if they had moved through transition seven – yes, transition *seven*, the move to the colonial living found in ants, bees and termites, no longer just transition *six* – yet without having any of the multiple additional chemical, physical, genetic or other biological mechanisms that are necessary for transition seven. The group-selectionists have realised that something very weird indeed does need explaining in human behaviour, although sociobiologists have similarly drawn a parallel with transition seven, as with: "we are more like ants or termites who live as slaves to their societies" (Ridley 1996, p.6).

When genetic group selection is sometimes argued to also apply to a few non-human species, or at least rare and occasional behaviours within a few non-human species, the results can invariably be explained more robustly, and more parsimoniously, as the operation of genic selection. In a later section we will be fully reviewing those rare instances where genetic group selection may seem to operate, but the point is they require unique conditions which are almost never realised in nature, such as where selection is of the maximum possible intensity. When it comes to what Williams will call "adaptive mechanisms worthy of the name", it is only with one single animal species that we are seeing something which genic selection has difficulty immediately explaining, albeit for the above reasons genetic group selection becomes an even less likely candidate. Because group size is not the only problem here, and genetic group selection becomes a more widely unworkable explanation of human behaviour, as Darwin, George Williams, and John Maynard Smith all argued. In fact, and for the record, neither of the last two saw a true intellectual distinction between the sociobiologists and this third group, always suggesting that human sociobiology was really just an even more incoherent and muddled form of genetic group selection. Ed Wilson would ironically decamp from his own discipline of human sociobiology to the genetic group-selectionists sometime between 2007 and 2010, thus highlighting to everyone who hadn't already noticed it the deep and abiding link between these two traditions.

Yet although we must still get to the bottom of the puzzle of how selection is really acting, the main takeaway for this first chapter is that *whatever* interpretation of evolutionary theory you cling to from out of these three – the only extant three – posited major traditions, the bottleneck really does appear to be happening after transition six, and all the traditions give us an extraordinary and even wholly unique eighth transition to explain intelligence, or at the very least human group size and cohesion. Logically of course Brian Cox might still be right, and there may be a second bottleneck sitting somewhere earlier on, be that his suggestion of transition four, or somewhere else. But this is where the principle of parsimony, sometimes called Occam's razor, comes in. In science the principle of parsimony says that you don't invoke two separate improbable events when just one will explain the situation. The simpler solution is more likely to be correct. We know there is a bottleneck somewhere, but as all the traditions are pointing to a truly major and extraordinarily unique bottleneck after transition six, at this stage we should not be looking for bottlenecks before transition six. They may still exist, of course, and life on Earth may have gone through more than one astonishingly improbable chokepoint, but this is now much less likely to be the answer, especially as transition four is not necessarily to be particularly unexpected.

Option	Requires	Conclusion?
Darwin / Genic selection	Second inheritance mechanism	Cosmic dinosaurs
Group selection	Move to group-selectionism	Cosmic dinosaurs
Human sociobiology	Reversal of the genic selection process	Cosmic dinosaurs

So transition four is not likely to be the bottleneck, and whatever interpretation of evolution by natural selection you hold to our galaxy appears to be teeming with dinosaurs. Millions of planets in just our galaxy may well have enormous and terrifying reptile-like beasts roaming the lands, and giant creatures swimming the seas. The Great Silence it may still be, but it is a great silence that seems now to come out of a vast background of screeches, howls, trumpets, bellows, and roars.

COSMIC DRAGONS, NO; COSMIC DINOSAURS, YES

We need to mention here that many scientists have considered the bottlenecks – chemical, physical, biological and cultural – that may be leading to the Great Silence. An associated term mentioned above, the Great Filter, comes from a 1998 paper by the George Mason University academic Robin Hanson. Hanson, now also a research associate at Oxford's Future of Humanity Institute, cited nine steps, including six of Maynard Smith and Szathmáry's evolutionary transitions, in the path from interstellar dust to intergalactic colonisation (Hanson, 1998). Nevertheless, like almost all who have looked at the possible filters, Hanson failed to note that Darwin had made a critical distinction between a single inheritance mechanism and a dual inheritance mechanism, and that Darwin was thus already giving us both cosmic dinosaurs and only two possible forms of deliberative extraterrestrial life, one of which would seek to exterminate us. And that Darwin was promising us an existential struggle between humanity and self-aware machine intelligence.

Although most of his nine steps were equally hard for Hanson ("overall, we might estimate a total of roughly seven to nine hard steps"), Hanson also touched on themes, including self-destruction, and galactic colonisation and dominance ("colonizing other stars and galaxies"), which have helped make the Great Silence of such interest to both techno-futurists and certain techno-billionaires. "Research into SETI and the evolution of life does much more than satisfy intellectual curiosity – it offers us uniquely long-term information about humanity's future", Hanson wrote. SETI is of course the Search for Extraterrestrial Intelligence, at first funded by NASA and the American taxpayer, and later a non-profit institute with trustees including the astronomer the late Carl Sagan. Hanson's Great Filter is sometimes quoted by techno-futurists and those keen to see parts of humanity spread beyond Earth so that we might never go extinct, and for example in May 2021 Elon Musk tweeted about it. "Becoming multiplanetary is one of the greatest filters", wrote Musk. "Only now, 4.5 billion years after Earth formed, is it possible. How long this window to reach Mars remains open is uncertain. Perhaps a long time, perhaps not. In case it is the latter, we should act now", referencing his own investment in eventual settlement on Mars. Musk was responding to a tweet drawing his attention to a November 2020 article in *Astronomy Magazine* that had asked whether Hanson's Great Filter was a possible solution to the Fermi paradox (Adler, 2020). Musk's "how long this window? … we should act now" message is an article of faith within large parts of the techno-futurist and transhumanist communities, because as Hanson put it: "the easier it was for life to evolve to our stage, the bleaker our future chances probably are".

We will consider later the work of some others currently writing on extraterrestrial intelligence, such as the Harvard astrophysicist Avi Loeb, and we will return to both the techno-futurists and the transhumanists when we deal with, respectively, artificial (or machine) intelligence and enhanced natural life, but before we move on from the evolution of complex life to the evolution of intelligent life, let us complete Darwin's first insight. Complex multicellular life can come in a near-infinity of different forms, which is what we can expect throughout the galaxy, but not actually an infinity of different forms. Billions of different species exist or have existed on just our planet, but there have always been constraints to what could have evolved. Physics and biochemistry place natural limits, even if these limits may change over time, but history also places constraints.

In his contribution to Orion Books' Science Masters series, *Plan & Purpose in Nature*, George Williams introduces us to what biologists call phylogenetic constraint, or phylogenetic inertia. "Evolution never designs anything from scratch. It can only tinker with whatever happens to be already there, saving those slight modifications that provide immediate benefits, culling those that cause harm. Much of anatomical human nature derives not from anything currently desirable but from adaptive changes made in the early history of the vertebrates" (1996, p.175). We have two pairs of limbs, wrote George, not for functional reasons but for purely historical ones. "The first lungfish that crawled from the water and pushed its way through the mud did so with the help of two pairs of appendages" (p.177).

Williams used to joke that phylogenetic constraint explains why you are unlikely to come across an angel; those celestial wings alongside their sword-wielding arms mean they would have a body plan involving three pairs of limbs, so couldn't have shared the same evolutionary history. And thus "in evolving wings, birds seriously constrained their evolutionary futures. They could never again use their forelimbs for terrestrial or arboreal locomotion or manipulation". Body plans can, however, change somewhat over time, and degeneration is another historical drama, although one usually played out at a different speed. Most bony fish have two sets of paired fins, but some have only one set, or even none, as with eels. Snakes are limbless – although some do have claw-like hind limb remnants – as their limbs were gradually selected to get smaller over time. Snakes have also lost eardrums and moveable eyelids, but gained vertebrae ("midline structures ... may be serially repeated any number of times in different species. The vertebrae themselves are a good example, ranging from a dozen to hundreds", pp.175–6). In general, degeneration can be enormously faster than generation. It required millions of years to make vertebrate eyes, but only thousands of years to lose them in caves.

What biologists term convergent evolution also highlights the constraints we must work within. Convergent evolution is the evolution in different lineages of similar structures, not because of a common ancestor, but through independent evolution under similar necessities and similar selective pressures. While there is some argument over the distinction between parallel evolution and convergent evolution, marsupials and placentals have certainly converged in body form and function across a number of ecological niches, such as the hyena and the Tasmanian devil, the groundhog and the wombat, and the giant anteater and the numbat anteater. Australia's spiny anteater is not a marsupial, as it is an egg-laying mammal, but is another example of evolution converging on a similar solution. Sabre-toothed predators have likewise evolved in multiple distinct lineages of mammals, though as Richard Dawkins explains, "the convergence is not total" (1986a, p.94), as developmental origins, history and evolutionary opportunity will always be somewhat different. In *The Blind Watchmaker*, Dawkins goes on to devote half a chapter to examples of convergent evolution, including echolocation, electrolocation, and periodical cicada eruptions.

We have discussed the role of historical constraint in body form, and the role of ecological niches, and we are aware of the arms races in evolution as animals evolve in response to one another. We currently share this planet with the largest animal that has ever lived, the blue whale, which touches 170 tonnes and 35 metres in length. The largest, but these are still gentle giants, unless you happen to be krill. The mainly plankton-feeding whale shark grows to about 16 metres in length, yet until just three and a half million years ago our oceans saw a shark, megalodon, measuring maybe up to 20 metres in length, with 18-centimetre teeth and a bite force that has been estimated to be ten times as powerful as a great white. Because temperature and atmospheric gases also affect animal and plant life. Megalodon is thought to have become extinct as the seas cooled, habitats and birthing locations shrank, and life at the bottom of the food chain collapsed leaving less available, or at least less accessible, prey.

Yet although we share the planet with the largest creature in history, we have never shared it with the giant reptiles that died out some 66 million years ago. Scientists continue to debate what allowed so much gigantism to coexist at the same time. While the sauropod dinosaurs were the largest land animals ever, and weighing up to maybe 75 tonnes, remembering that they had to support their own weight, unlike the blue whale and megalodon, just the average weight range for these dinosaurs was a whole order of magnitude larger than that for the largest herbivorous mammals. The herbivorous sauropods were then tracked in body size by the carnivorous theropods, and as Sander *et al.* note, "the theropods were an astounding 12 times heavier than predicted by the regression equations for extant ectotherms" (2011, p.123).

We mentioned arms races, and there is an argument the herbivorous sauropods grew larger to avoid being preyed upon, while the carnivorous dinosaurs then needed to become larger to take down such giants. However, that could only ever be part of the answer. Until relatively recently researchers had hypothesised that dinosaur gigantism may have had something to do with a higher oxygen atmosphere, or greater plant productivity through changes in carbon dioxide, or higher ambient temperatures, but thinking has changed. Sander *et al.* (2011) recognised the work of others confirming that when oxygen levels fell in the Permian and Triassic, dragonflies decreased in maximum body size, and that in the current 21% oxygen atmosphere dragonfly wingspan is restricted to 16 centimetres due to design limitations of their tracheal respiratory system, whereas in the 30% oxygen atmosphere of the Carboniferous the same design permitted wingspans of over 70 centimetres in some of the extinct Meganisoptera. However, they found no evidence that resource availability and global physicochemical parameters were different enough in the Mesozoic to have led to sauropod gigantism. Oxygen levels actually seem to have been lower in the Jurassic and Cretaceous than today, while higher carbon dioxide contents were not constant enough to provide an explanation. Instead, the authors have suggested that evolutionary innovations, the most important being the very long necks which allowed more energy-efficient food uptake, sat alongside a specific phylogenetic heritage. A long neck could evolve because of a small head where food was ingested without mastication, coupled with the inheritance of hollow neck bones and an avian-style flow-through respiratory system (bit of connected trivia: it would be Darwin's bulldog, Thomas Huxley, who one hundred years ahead of his time would work out that birds had evolved from dinosaurs). Alongside other researchers, Sander *et al.* suggest that an oviparous mode of reproduction was another major contributor to sauropod gigantism, while Sander (2013) has updated the analysis including highlighting the importance of hatchling-adult body size difference. Nonetheless, while we might not have shared the planet with the giant reptiles that lived when the dinosaurs dominated, we do share it with the crocodiles, whose close cousins even out-competed the early dinosaurs. Some of the crocodile-line archosaurs were the apex predators of their period and reached almost 10 metres in length, and ruled until the Triassic-Jurassic mass extinction two hundred and one million years ago killed off much of this group, leaving the dinosaurs to inherit the land, and begin their growth in size.

So where does this leave us? A recent analysis of exoplanet data collected by the Kepler space telescope, and the first to use the full Kepler dataset, has suggested that even using the most conservative estimate there are a minimum of 300 million sun-like stars in our galaxy with at least one habitable rocky planet, and that the average expected number is over two billion habitable rocky planets. Evolution takes time, and the major transitions within evolution require both time and a degree of physical stability, but ours is an old galaxy, and parts of it very old indeed. Building from Darwin's realisation that the ultimate bottleneck must exist at the level of deliberative life, we can expect the galaxy to be packed with complex life of almost every conceivable size and body plan, limited only by developmental history, ecological necessity, and the laws of chemistry and physics. Giant dinosaur-like creatures may exist on millions of those planets, and because they can. But there are limits on what should be expected and, for example, giant, flying, fire-breathing creatures are a little unlikely ever to have evolved. As the humourist Terry Pratchett wryly noted, not only would a dragon burn its own lips off and likely never get off the ground, but the destiny of any young fire-breathing dragon would be to live fast and die wide, because with an inside like an alchemist's lab one misplaced hiccup and it would find itself plastered across the landscape. Evolutionary opportunities are severely limited for creatures that have a tendency to explode or self-immolate.

For Darwin, then, a reluctant "no" to cosmic dragons. But an enthusiastic "yes" to cosmic dinosaurs, and to planets with enormous and deadly marine creatures with 18 centimetre teeth, and crocodiles the size of buses. More than that, though; "yes" to a universe absolutely teeming with complex multicellular life under each of the three extant traditions of evolutionary biology: gene-selectionist, group-selectionist and sociobiologist. Because this forms the basis for the extraordinary realisation that runs throughout this book. All three traditions are ultimately promising us cosmic dinosaurs, but all three traditions are also going to offer us a level of deeply hostile extraterrestrial intelligence, and all three traditions are going to guarantee us machine intelligence that will mean an existential struggle for survival.

"I fully admit that it is the highest & most interesting problem for the naturalist", Darwin wrote in December 1857 to Alfred Russel Wallace, the co-developer of the theory of evolution by natural selection (Raby 2001, p.134). Darwin and Wallace were corresponding here about the human animal, though neither Darwin nor Wallace were precious about our species, and for both of them humankind was just another animal species among so many tens of thousands. In fact, Wallace, "the ever-vigilant defender of natural selection, the ultra-adaptationist, the most Darwinian of Darwinians" as he has been called (Cronin 1991, p.353), would later term Darwin's highest and most interesting dilemma "the inverse problem" for the naturalist, and in an 1870 work entitled *The Limits of Natural Selection as Applied to Man.* "In order to account for facts which, according to the theory of natural selection, ought not to happen" (reprinted in Wallace 1891, p.188).

So why if they were not being precious about humankind would the two fathers of evolutionary biology be here singling out just one particular species as "the highest & most interesting" problem in evolution, and as both the intellectual and behavioural archetype that "ought not to happen"? They were doing so for the same underlying reason that goes on to give us cosmic dinosaurs, murderous aliens, and genocidal machine intelligence, and however we present our evolutionary theory; gene-selectionist, group-selectionist, or sociobiologist. High human intelligence, very large and peacefully cohesive human groups, and human morality; none of this was even remotely to be expected by Darwin and Wallace under the rules of their own theory. Because none of this is predicted by, is explained by, what we can today call transition six, or even transition seven, both of which were effectively known to Darwin and Wallace. Humans do not behave as we should under the rules of transition six, or the rules of transition seven, so for Darwin and Wallace, and for any exponent of modern evolutionary theory, there has to be some further transition, coming with all those extraordinary and deeply worrying implications mentioned above.

Saying that, this book will still have to come down on the side of just one of these three evolutionary traditions, and despite the fact that each of the three produces implications, particularly as regards artificial intelligence, that are far too important to ignore. The book will be trying to present where it can the conclusions for all three traditions next to each other, because each has at least earned the right to consideration and some respect. All three deserve a degree of admiration because all three are trying to justify intellectually why humankind really is the highest and most interesting problem in evolution, and please understand that it has nothing to do with flattering our species. All three are ultimately juggling with that key eighth transition.

So all three are working with an understanding that has been almost wholly missed by other disciplines, which is to their credit, but the accommodation must stop at some point, and we must settle for one out of the three. Because not only can just one of the three be correct (and, judiciously, one of the three must be correct), but the reason that other disciplines have largely missed the importance of transition eight is that the struggle – the often quite vicious and nasty struggle – between the three traditions has regrettably shielded this final transition from general sight. Plus at least two of the three traditions have been so busy fighting to win on transition eight that they failed to even understand the wider consequences of their own arguments, while simultaneously shielding from view the vastly important implications for both multicellular life away from this planet, and any possible form of contemplative life, biological or artificial, and including our own species.

A final couple of points before we move into the rest of the book. We are not going to be interested in SETI, the Search for Extraterrestrial Intelligence, or more than lightly touch upon a barely glimpsed space rock that became an international sensation in late 2018 when it was suggested it might hold implications for extraterrestrial intelligence, and even though such a proposal required a huge degree of faith and wishful thinking. And we will simply brush past a July 2015 open letter signed by some of the world's leading scientists and technologists seeking to prevent the use of autonomous AI in warfare. Instead, we are only going to be concerned with the application – the first-ever such application, the only such application – of reason and logic to Darwin's key eighth evolutionary transition, where the implications for both natural (thus now meaning terrestrial and non-terrestrial) intelligence and artificial intelligence are wondrous, stark, and undeniable, at least insofar as they resolve across the three different biological traditions. Where we get to start with cosmic dinosaurs, but then move on to so much more.

Yet this then necessarily becomes about much more than just one further transition, as it is seemingly also *the completion of the entire Darwinian project.* Until we openly recognise this inverse problem, until we explicitly address Darwin's highest and most interesting problem for the naturalist, the evolutionary project will remain incomplete. The eighth transition problem unaddressed, Darwinism can still coherently explain body size and body form, from smallest to largest, from cold-bloodedness to warm-bloodedness, from fins to feathers. However, the eighth transition problem unaddressed, Darwinism cannot explain the full range of intelligence, not human-level intelligence, not extraterrestrial intelligence, and not the realities of artificial intelligence. Unaddressed, Darwinism cannot explain human group size and interaction, cannot explain terrestrial civilisation, cannot explain non-terrestrial civilisation, and cannot explain the coming AI experience. The eighth transition problem unaddressed, the Darwinian project remains incomplete.

All three biological traditions offer us different answers to this "highest & most interesting problem", but this need for a further transition – be that a second inheritance mechanism, be that natural selection operating at a fundamentally different and unexpected level, or be that new mechanisms of evolution never needed in the previous billion years – is an issue the public is wholly unaware even exists. In a sense it becomes more immediately critical to actually take the time to explain the underlying problem than it perhaps is to provide the correct answer to the problem, at least at this stage. And if this sounds a little trite, do understand that providing an answer without explaining the problem has been the *modus operandi* to date. As just one key example, in *The Selfish Gene* Richard Dawkins devoted an entire chapter to answering Darwin's highest and most interesting problem, firmly coming down on the side of Darwin's solution of a second inheritance mechanism, or what Dawkins termed a new replicator. But this chapter was his most controversial chapter, misunderstood because it made no serious attempt to explain the underlying problem he was trying to solve. From Dawkins there was no talk of transitions in evolution, no explanation that the chapter was made necessary because human-level intelligence and human group harmony could not be explained by the earlier transition to multicellular life, or even by the natural world's move to eusociality. No pointing out that all extant biological traditions – from gene-selectionist, to group-selectionist, to sociobiologist – are seeking the answer to this further transition problem. No highlighting that in a very real sense the only distinction between the three evolutionary traditions is in their answer to the eighth transition problem. Yet Dawkins was then criticised for this chapter by biologists like Ed Wilson, David Sloan Wilson and Robert Trivers, who were themselves providing *without explanation and largely without awareness* different versions of Darwin's eighth transition.

The "highest & most interesting problem for the naturalist", and however it ultimately be resolved, is then the particular problem we at least need to comprehend before we should attempt to predict the probability of large animal life off this planet, and it is certainly necessary to understand it if we are interested in the components of human intelligence, the faces of non-terrestrial intelligence, and the character of artificial intelligence. Further, though, it is this eighth transition problem that we need to appreciate before we can attempt to really answer the question of the levels natural selection can and cannot operate at, and the particular mechanisms that are available to natural selection; it is thus the singular problem to conjure with if we ever hope to complete the Darwinian project.

2

DARWIN "LETS US DOWN". REALLY?

"How did we evolve from being merely social to being moral? His analysis begins promisingly, ... but when Darwin turns to his other solution, he lets us down."

— **Helena Cronin**, sociobiologist (1991, pp.326–7)

In the first chapter we set out the evolutionary deductions for cosmic multicellular life without fully needing to separate the main interpretations of natural selection, and because all of the main traditions converge on something very unusual happening after transitions six and seven. All of the main traditions give us an astoundingly rare eighth transition that acts as the evolutionary bottleneck, even if they are offering different interpretations of that eighth transition. So, all of the main traditions provide us with cosmic dinosaurs as our starting point. But to fully understand the further implications for both natural (non-terrestrial, but also terrestrial) intelligence and machine intelligence, and as possible across transitions six, seven, and eight, we must now pull apart the differences between the three major extant evolutionary traditions, being both Darwin's explanation of intelligent life and the two interpretations opposing Darwin's explanation of intelligent life. Before we go any further we must briefly delve into the history, and theory, of evolution by natural selection.

In 1859 in *On the Origin of Species* Darwin wrote that "natural selection can act only through and for the good of each being. ... Natural selection ... will adapt the structure of each individual for the benefit of the community; if each in consequence profits by the selected change" (pp.133–5). Darwin was what is called an individual-selectionist; he believed that nature could select only through and for the good of the individual. Darwin here is explicitly disavowing community or "group" selection, or the idea that nature will select for the benefit of the community or group rather than its component parts. Here any benefit to a group can only be as a by-product of what is good for its individual members.

Evolution – the understanding that species change with time to form entirely new species because of the inheritance of different characteristics – was not new when Charles Darwin produced his work. Transmutation, as it was then called, was already in the air when Darwin published *Origin of Species*. Charles' grandfather, Erasmus Darwin, was among those who had produced early theories of "perpetual transformations" within nature. The Frenchman Jean-Baptiste Lamarck had proposed an even more influential, and alternative, theory of evolution that Darwin himself toyed with for many years before ultimately rejecting. And Robert Chambers' bestselling *Vestiges of the Natural History of Creation* had already prepared Victorian society for the fundamental shift in world views, as the old biblical static ordering of separately created species would be swept away, to be replaced by a transmutationary landscape. But what Charles Darwin and his contemporary Alfred Russel Wallace were to discover was one particular and ultimately very convincing explanation for evolution, that of evolution by natural selection. Wallace, by the way, remains the great footnote in Darwinian history; the shy, asthmatic, often penniless surveyor-turned-naturalist who, though little remembered today outside of scientific circles, independently developed a remarkably similar theory of evolution by natural selection at a time when Darwin was still sitting on his own explosive conclusions. It was Wallace who forced Darwin to publish, and their papers were then presented together at the Linnean Society in July 1858. It was Wallace who fired the starting gun on so much of modern intellectual life.

Natural selection applies to entities with the characteristics of multiplication, variation and heredity. What this means is that natural selection works on entities that can make copies of themselves. Variation means that copying, however, will never be perfect, and it is random copying errors that are essential if the process of evolution is to occur. Once the error or mutation has occurred, heredity will ensure that a mutation can reappear in all future generations. Over time evolution will tend to favour higher multiplication, longevity (the ability to survive long enough so as to be able to reproduce) and heredity. Natural selection is the process that determines that traits which are conducive to superior replication – Darwinian fitness – will, of mathematical necessity, tend to become increasingly common in a population over time.

Adaptation was key to Darwin's explanation of evolution by natural selection. Adaptation, or perfection of design, is explained as the gradual adjustment in form and behaviour as selection saves what is useful, and discards what is less useful, and slowly improves designs to fit an environment. Natural selection can this way often produce extremely precise, though never perfect, contrivances. Gradual saving of slight improvements will, over very many generations, allow a blind, physical process of accumulating small beneficial mutations to create the wonders of nature that the early nineteenth-century theologian William Paley had seen as proof of God's intervention; the eye or the hand. Paley's assumption of a forward-looking Creator necessary to fashion the most intricate designs in nature – natural equivalents, said Paley in 1802, of such perfect human creations as the watch – falls away. The "blind watchmaker" of natural selection becomes the main mechanism for creating the wonderful diversity of plant and animal life we see around us.

Even during Darwin's time, though, there had been distracting, and sometimes deeply worrying, objections to natural selection, such as when the brilliant but self-important physicist Sir William Thomson, later Lord Kelvin, mistakenly challenged that the Earth was simply not old enough to allow for Darwinian gradualism. But Kelvin, whose name would be given to the temperature scale, had miscalculated by a factor of almost fifty the Earth's age, estimated through the relative rate of heat loss, as he knew about thermal conductivity, but radioactivity and nuclear processes were at that stage unknown. Darwin also did not have a theory of inheritance, and a legitimate objection was that without a robust theory of discrete inheritance beneficial variations would have blended away, or been swamped, under natural selection, not carried forward. The period from the 1880s up until the 1920s has been called "the eclipse of Darwinism", and where evolution by natural selection was seen by biologists as just one theory among many, and not a particularly good candidate at that. Real challenges remained for Darwin and Wallace's theory until it was synthesised with Gregor Mendel's rediscovered work on peas and particulate, not blended, inheritance in the early twentieth century. From the 1920s, and after more rigorously mathematical approaches and work including Thomas Hunt Morgan and his famous fruit flies with their discrete mutations, geneticists were comfortable combining population-based Mendelian genetics and Darwinian gradualist natural selection. Neo-Darwinism had now given way to what by the 1940s was known as the modern synthesis of Darwin and Mendel, and it would sweep away all competition.

Yet even outside of – at least at the time – seemingly legitimate scientific concerns, there have always been attempts to soften natural selection when it comes to our species, to argue that other animals might have evolved under the normal rules, but that humans got to step outside these processes. Sir Richard Owen, today perhaps best remembered for coining the term "dinosaur", was a palaeontologist, a naturalist, and a skilled comparative anatomist. But while Owen did not deny that humankind was an evolved primate, he detested Darwinian gradualism, and saw instead the first human as having sprung fully formed from the womb of an ape ancestor. Owen had claimed that "man should stand in a special sub-class, one reserved for him alone" (Desmond & Moore 1991, p.453), because humanity had special attributes that must have been bequeathed us by our Creator, and held that the human brain possessed a unique lobe, the hippocampus minor, not found in a gorilla's brain. Thomas Huxley, though, soon joined by others, would clash with Owen throughout the first years of the 1860s as their dissections of gorilla, chimp and orang-utan brains would reveal the elusive hippocampus. Owen was the founder of London's Natural History Museum, where his statue sat in the main hall until 2009, when in a move that would have had Owen turning in his grave it was replaced with a statue of Darwin.

But as it was with Owen, so it was with many other great establishment scientists, including Lord Kelvin. Kelvin simply did not like the implications of Darwinian theory, and he and Huxley fought tooth and nail over the evidence; "this method of treating my 'case' is perfectly fair, according to the judicial precedents upon which Professor Huxley professedly founds his pleading", he commented in 1869 (reprinted in Kelvin 1894, p.88). But Kelvin, like Owen, was not so much hostile to evolution as to evolution by natural selection at the level of the individual, to Darwinian gradualism, and to anything which did not make humankind a reverent and special case. Kelvin admitted that he sympathised with the general idea of evolution, but he could not accept the particular mechanism proposed by Darwin and Wallace. Kelvin agreed with Sir John Herschel's objection that natural selection is too much like the Laputan method of making books, in *Gulliver's Travels*, by the random mechanical combination of words, "and that it did not sufficiently take into account a continually guiding and controlling intelligence" (Brush 1982, p.13). Herschel, the celebrated astronomer and mathematician, was at first a strong positive influence on Darwin, including with the two meeting in 1836 at the Cape of Good Hope. As Darwin commented, "Sir J. Herschel's *Introduction to the Study of Natural Philosophy*, stirred up in me a burning zeal to add even the most humble contribution to the noble structure of Natural Science" (Warner 2009, p.432). Herschel was a hero to Darwin, and Darwin was therefore stung by Herschel's criticism of the theory of natural selection. "I have heard by round about channel that Herschel says my Book 'is the law of higgledy-pigglety'. What this exactly means I do not know, but it is evidently very contemptuous. If true this is a great blow & discouragement" (p.438). Herschel failed to accept the essence of Darwin's proposed mechanism; "Herschel believed that *directed* variations were necessary". Herschel's book, *Physical Geography*, would go on to emphasise the need for "an intelligence, guided by a purpose" that must be continually in action "to bias the directions" of the evolutionary changes.

Yet even without the flawed but distracting objections of Owen, Herschel and Kelvin, Darwin knew that for very good reasons his theory of evolution by natural selection could not account for either morality or large group close cohesion within a mammalian population. Darwin understood the struggle for existence, and his theory explained aggression, but also sociality and co-operation. Yet, mathematically, co-operation had to be severely limited in mammals, restricted to what the sociobiologist Helena Cronin above describes as the "merely social". Darwin was aware of the competition and the bloodletting where a newly dominant young male gorilla "killing and driving out the others, establishes himself", as he wrote in *Descent of Man* (1871, Pt. ii, p.363). He knew about baboons using weapons to attack other baboons, where they would roll down great stones, and about orang-utans using sticks and fruit as missiles, although he also appreciated that primates could collaborate in warfare. But natural selection could not explain the evolution of human sentiments like patriotism, justice, fair play and virtue. The numbers simply wouldn't work for Darwin. To explain both human morality and human group cohesion Darwin needed to turn to another answer outside of biological adaptation; as we shall see, Darwin needed a second inheritance mechanism.

SELECTION AT THE LEVEL OF THE GENE

The selfish gene revolution – the modern biologists' explanation of Darwinism that sees selection as predominantly operating at the level of the smallest unit, the gene, rather than at the level of the individual – is, as noted, generally held to have started in 1966 with the publication of George C. Williams' *Adaptation and Natural Selection*. Williams saw himself as continuing in a tradition first anticipated by major biologists such as R.A. Fisher, J.B.S. Haldane and Sewall Wright in work dating back to the early 1930s, but it was not until Williams' clear analysis that the incompatible traditions began to diverge.

Gene-selectionism sees apparent individual selection reinterpreted as not what is good for an individual but as what is good for its genes. Individual selection is often for practical purposes synonymous with gene selection, and unless evidence contradicted the gene was to be recognised as the fundamental unit of selection, and Williams proposed that in evolutionary theory a gene could be regarded as any "hereditary information for which there is a favorable or unfavorable selection bias equal to several or many times its rate of endogenous change" (1966, p.25). This concept of the gene as any unit of developmental information visible to natural selection has become the standard definition of the gene within the discipline, and the start of what is called the adaptationist programme within Darwinism. However, the terms selfish gene and selfish gene-ery did not emerge until Richard Dawkins popularised the programme a decade later in his 1976 bestseller, where he explained that "I must argue for my belief that the best way to look at evolution is in terms of selection occurring at the lowest level of all. In this belief I am heavily influenced by G.C. Williams's great book" (p.11).

Gene-centred natural selection has swept across biology as the orthodox modern interpretation of how evolution overwhelmingly acts. But gene-centred evolution has to obey certain rules, just like Darwin's individual-centred evolution has to. At least within biology, gene-centred evolution has largely knocked aside earlier interpretations such as the idea of selection acting with major effects at the level of the group or species (group-selectionism), which is prone to problems like subversion from within, somewhat similar to what is known in economics as the free rider problem. But Darwin's mathematical problem that Williams, John Maynard Smith, Bill Hamilton and George Price would each rediscover, and that Richard Dawkins would also quietly highlight in the late 1970s, was that – irrespective of whether natural selection operates at the level of the individual as Darwin believed, or largely at the level of the gene as Williams and Maynard Smith had shown – biological evolution could not explain a well-developed moral sense or capacity for fair play, and for key evolutionary reasons. Chimpanzees live in "a world without compassion", writes the celebrated primatologist Frans de Waal (1996, p.83). And chimpanzees live in this world without compassion because of the biological rules discovered by Darwin and Wallace, and refined by Williams, Maynard Smith, Hamilton and Price.

SELECTION AT THE LEVEL OF THE GROUP

Williams' 1966 book, subtitled *A Critique of Some Current Evolutionary Thought*, was also a criticism of theories of group selection, or the enduring idea that nature might select for the advantage of the group even at the cost to the individual. Biologists had a long track record of appealing, sometimes only implicitly, to "greater good-ism" and the idea that nature may select for the good of the group, local population or species. Mutually beneficial co-operation and reciprocity had been strongly debated in late nineteenth-century Darwinism, especially after the Russian naturalist and anarcho-communist Peter Kropotkin published his thoughts on "mutual aid". Kropotkin wanted to put benevolence back into nature, where Darwin had seen competition and bloodletting, and Kropotkin was initially writing against Huxley's views on the struggle for existence, and the Hobbesian war of each against all. "But neither Rousseau's optimism nor Mr. Huxley's pessimism can be accepted as an impartial interpretation of nature. … The ants and termites have renounced the 'Hobbesian war,' and they are the better for it", wrote Kropotkin (1890, pp.339, 344). Kropotkin theorised a benevolent instinct – "a feeling infinitely wider than love or personal sympathy", as he put it in the introduction to a 1902 longer exposition on mutual aid – and Harvard's Stephen Jay Gould, in his essay *Kropotkin Was No Crackpot*, explains that for Kropotkin co-operation "must balance or even predominate over competition", and that "Kropotkin sometimes speaks of mutual aid as selected for the benefit of entire populations or species" (1988).

Stephen Jay Gould is one of the best-known names in late twentieth-century genetic group-selectionism, and here he contrasts Kropotkin's mutual aid with Darwin and Huxley's "'gladiatorial' view of natural selection", noting that Kropotkin is there for those "who do wish to find a basis for morality in nature and evolution". Now whether one sees Kropotkin as a fully fledged group-selectionist, or as just advancing parallel ideas that John Maynard Smith explained time and again could be subverted from within, in 1962 in his *Animal Dispersion in Relation to Social Behaviour* the English biologist V.C. Wynne-Edwards (who coined the term group selection, a term which therefore hadn't existed in Kropotkin's time) would propose that nature could and did select for the benefit of the group even at the expense of the individual. A serious attempt to resolve the group selection question was now unavoidable, and it fell to both Williams and Maynard Smith.

Williams' book-length response was a review of the range of instances cited as examples of group selection. While noting that group selection was not theoretically impossible, Williams concluded that the adaptations in question could almost invariably be explained in terms of selection at levels lower than the group. Furthermore, Williams pointed out in *Adaptation and Natural Selection* that group selection failed science's parsimony test; theorists did not need to appeal to a higher-level solution when a lower-level solution had already solved the problem. Williams did not argue that group selection could not happen, only that it would usually be too weak to produce noteworthy effects, and because of the much stronger and often counteracting forces operating at the lower levels. Williams freely admitted that "actually I think group selection needs much more attention" (pers. comm., 2 September 2003), and it was his willingness to listen to alternative ideas – even ideas that he had made his reputation by undermining – that left him on good terms with almost everybody in the field.

Maynard Smith's analysis in the journal *Nature* in 1964 was to similarly note the difficulties with group-selectionist theories, including the damaging spread of the "'anti-social' mutations" that Wynne-Edwards was denying would propagate. "Every time a group possessing the socially desirable characteristic is 'infected' by a gene for anti-social behaviour, that gene is likely to spread through the group", Maynard Smith wrote. "Thus it would only be plausible to suggest that there are genetic reasons why anti-social behaviour should not increase if it were also suggested that selection had already produced an extreme degree of anti-social behaviour, and this is precisely what Wynne-Edwards denies. In fact, 'anti-social' mutations will occur, and any plausible model of group selection must explain why they do not spread" (pp.1145–6). George Williams would make similar points in his 1966 work, preferring the term "poisonous" mutations to "anti-social" mutations. "Only one locus is involved. One cannot argue from this example that group selection would be effective in producing a complex adaptation involving closely adjusted gene frequencies at a large number of loci. Group selection in this example cannot maintain very low frequencies of the biotically deleterious gene in a population because even a single heterozygous male immigrant can rapidly 'poison' the gene pool" (1966, p.118).

Evolution is a struggle for survival and works because it is. Organisms within groups are still competing against one another, and the organisms that begin to take the benefits without paying the costs will be favoured by evolution. Those organisms that acted for the good of the group, rather than at all times being driven by what is good for themselves and their genes, would be taken advantage of and would not be the successful ones. Group selection would explain morality and fair play *if* it were a coherent and influential force in biological evolution. Unfortunately, the theoretical and physical evidence is that it is not a meaningful force. Williams himself was not averse to considering effects from group selection, but while not denying that group selection could occur it was "usually too weak to produce noteworthy effects" (letter, 3 December 1998). "Group selection needs much more attention, but not with the stupidities published by people like Steve Gould and others who discuss 'species selection'. Surely any separate clades, subject to extinction, can be subject to selection. ... My personal friend and theological enemy David Sloan Wilson ... champions a locally benign form of group selection that enhances the collective fitness of large groups of organisms. I think this kind of selection unimportant" (letter, 2 September 2003).

What gene– and individual-selectionists argue is that group selection is likely to be a very weak force in evolution because the conditions required are so onerous. A group consisting almost entirely of altruists *can* do better than a group consisting entirely of selfish individuals, because the altruism benefits the entire group. The selfish group won't have this advantage. But, as Williams and Maynard Smith demonstrated mathematically in the mid 1960s, the problem is explaining how a group consisting entirely of altruists could have come about in the first place. "There is one special form of group selection which is worth considering in more detail, because it can, perhaps, explain the evolution of 'self-sterilizing' behaviour. ... With an intermediate amount of gene flow between colonies, selection could both establish and maintain timid or altruistic behaviour, provided that colonies with altruistic behaviour have a large selective advantage, and that colonies are founded by very few individuals. The model is too artificial to be worth pursuing further" (Maynard Smith 1964, p.1146).

"Lewontin has produced what seems to me to be the only convincing evidence for the operation of group selection", wrote Williams back in 1966 (p.117).[2] The evidence here referenced

---

[2] Williams was writing this in the mid 1960s, and once we began corresponding in the late 1990s he pointed out that D.S. Wilson and others had since provided evidence of group selection producing female-biased sex ratios in species with special population structures, and that others, including himself, had been saying that parasitic virulence will reflect the balance of within-host and between-host selection.

came from two Lewontin papers in the early 1960s demonstrating the far higher than expected frequencies of t-alleles in the American house mouse, which were in the highly unusual situation of seemingly being maintained in populations, as although the mutant allele produces sterility, extinction is at a higher rate in smaller groups. "It should be emphasized that this example relates to genes characterized by lethality or sterility and extremely marked segregation distortions", continued Williams. Selection of such genes "is of the maximum possible intensity", and hence "I question only the effectiveness of this extinction-bias in the production and maintenance of any adaptive mechanisms worthy of the name" (pp.117–9). Or as Richard Dawkins and the Oxford zoologist Mark Ridley put it, "mathematical models show that, except in very special conditions which are almost never realized in nature, group selection will lose because it is so slow" (1981, p.22). In a group consisting of both altruists and selfish individuals the altruists will be eliminated by within-group selection because of the problem of subversion from within. It is theoretically possible that a wholly altruistic group could arise if the group was small in number. In such a case, random genetic drift could hypothetically establish altruism within such a population. However, it could only be immediately maintained if there is little migration in from normal selfish groups; otherwise, the new immigrants would end up driving the altruists to extinction.

For group selection to be a serious force in evolution it would require between-group selection – where the altruistic inclination can be favoured because it would favour the group in competition with other groups – to be stronger than any within-group selection against altruists. Building from his own work in the 1960s, in the 1970s Bill Hamilton took the theories of Williams and Maynard Smith further to show that the force for group selection would only be able to continue indefinitely if there was periodic re-assortment of the various groups such that altruists could be re-concentrated in some groups. If this didn't happen such altruists would inevitably be eliminated by within-group selection. But the problem is coming up with a process of re-assortment that allows altruists to recombine in such a way that they will not immediately be taken advantage of by non-altruists. "I shall argue that lower levels of selection are inherently more powerful than higher levels ... It reveals a group-selection component which is not zero but which is bound in an unchanging subordination to the individual selection component" (Hamilton 1975, pp.330, 335).

While the group-selectionists tend to see it differently, orthodox Darwinism has since been a continuously improving attempt to explain natural selection as operating at the lowest levels, that which is of benefit to the individual or, more specifically, its genes. But even the group-selectionists understand there is a very real perception problem with hypothesising the evolution of morality. Recall Frans de Waal, the primatologist who told us that chimpanzees live in "a world without compassion". De Waal is openly group-selectionist, but he himself tells us about Mozu, an oft-filmed female snow monkey, or Japanese macaque, who had survived in a Japanese National Park despite severe deformities. Because even de Waal, usually so hostile to gene-selectionism, must admit that "there is no shred of evidence that other monkeys have ever gone out of their way to assist her in her monumental struggle for existence" (1996, p.7).

SELECTION AT THE LEVEL OF ... ERM

In 1975 the Harvard entomologist Edward O. Wilson, a recognised expert on insect behaviour, was to coin a new term: sociobiology. Sociobiology was defined as the scientific study of the biological basis of all forms of social behaviour in all kinds of organisms, including humans. This field of study extended to humankind could have been non-controversial; there logically *is* a biological basis to study, be it that the final answer is 100%, 0%, or a shifting range somewhere in between. Richard Dawkins made the same point in his essay "Sociobiology: The New Storm in a Teacup": it is a field of study, not a point of view, Dawkins explained (1986). The problem, though, was never the field of study; the problem was that central to sociobiology from day one was the claim for "morality as a biological adaptation", central to sociobiology was Bob Trivers' suggestion that when it came to human evolution nature simply broke the billion-year pattern. Exactly as with Gould, Kropotkin and genetic group selection, sociobiology was to become another intellectual home for those "who do wish to find a basis for morality in nature and evolution".

In his 700-page magnum opus *Sociobiology* Wilson summarised the previous five decades of research into animal social behaviour by biologists all over the world. But in the final chapter Wilson turned to humankind, and provided his own rather more unique interpretation; as Maynard Smith put it: "the last chapter – it seemed to me half-baked, silly" (in interview with Segerstrale 2000, p.241). It was half-baked for Maynard Smith because not only did Wilson wish to put cultural behavioural differences down to biology, and thus develop a new discipline of "anthropological genetics" focusing on cultural behaviours, but because Wilson admitted that demonstrating "the *genetic evolution of ethics*" (emphasis in the original) was a "missing" but "important piece" of the sociobiology project. And notwithstanding the problem that "to the extent that unilaterally altruistic genes have been established in the population by group selection, they will be opposed by allelomorphs favored by individual selection" (1975, p.563).

While not offering any evidence for "the *genetic evolution of ethics*" in 1975, this didn't stop Wilson suggesting to the *New York Times Magazine* that year that "Human Decency is Animal"

(1975a). And three years later he would publish a book solely about human sociobiology, *On Human Nature*, where he would explicitly claim that "morality evolved as instinct" (1978, p.5). Or as another early sociobiologist, Michael Ruse, put it: "The position of the modern evolutionist, therefore, is that humans have an awareness of morality – a sense of right and wrong and a feeling of obligation to be thus governed – because such an awareness is of biological worth. Morality is a biological adaptation. ... Perhaps we really ought to hate our neighbours, but we, poor fools, think otherwise!" (reprinted in 1989, pp.262, 271).

But when trying to explain this supposition Wilson was to write that where such genuinely altruistic "behavior exists, it is likely to have evolved through kin selection or natural selection operating on entire, competing family or tribal units" (1978, p.155), and that "the genetic capacity for blind conformity spreads" (p.187). The problem with the above is that the former claim is explicitly group-selectionist when claiming selection at the level of the competing (non-familial) tribal unit, while the latter claim is implicitly group-selectionist, as individuals without the putative trait of blind conformity would be taken advantage of and out-competed by their less helpful conspecifics. John Maynard Smith, one of the twentieth century's leading mathematical biologists, commented that "a few years ago, I worked through the equations in Lumsden and Wilson's *Genes, Mind, and Culture* and found them to be badly flawed" (reprinted in 1992, p.91). He wrote that he spent "several months trying to understand the maths" behind Wilson's 1981 collaboration with Charles Lumsden, a Canadian biologist, before determining that their mathematical models "fail to demonstrate any synergistic effect" (p.52). The sociobiologist Michael Ruse did point out quite early on that Ed Wilson had an obvious weakness for the explicitly group-selectionist argument – "curiously, the one human sociobiologist who is prepared to take seriously non-individualistic mechanisms of change is Wilson" (reprinted in 1989, p.163) – though Ruse failed to point out the only slightly less obvious implicit group-selectionism in his (see below) and others' work.

So then there is Robert L. Trivers, the Harvard postgraduate who first resurrected modern sociobiology in 1971, and who shares with Wilson the need to put human morality as a biological adaptation. "We routinely share food, we help the sick, the wounded and the very young", Trivers wrote under a section headed "Reciprocal Altruism in Human Evolution" (1985, p.386). Reciprocal altruism, by the way, is the exchanging of altruistic favours such that each benefits more from co-operating than from not co-operating. Because it was initially the American *Quarterly Review of Biology* that allowed Trivers to claim that "there has apparently been strong selection for a very complex system regulating altruistic behaviour", and that "selection may favor a multiparty altruistic system in which altruistic acts are dispensed freely among more than two individuals" (pp.39, 52). Where we are "acting altruistically toward a third individual uninvolved in the initial interaction" (p.53). All of which would be subject to Maynard Smith's subversion from within, all of which "will be opposed by allelomorphs favored by individual selection", and all of which sits outside the models of Darwin, Williams, Maynard Smith and Hamilton. Trivers noted that "there is no direct evidence regarding the degree of reciprocal altruism practiced during human evolution nor its genetic basis today, but … it is reasonable to assume that it has been an important factor in recent human evolution and … [has] important genetic components. To assume as much allows a number of predictions" (p.48). So Darwin's one hundred-year understanding of human-evolved altruism *as going no further than all other animal species* was here being set aside by a postgraduate's paper in an American scientific journal admitting no direct evidence and just assuming genetic components. Trivers commented that "the human altruistic system is a sensitive, unstable one". Well, that just about covers all the bases. Sensitive, unstable, no direct evidence, and no genetic basis. And the rest of the paper, the final seven pages of "assumptions" and "predictions", contained not one mathematical equation, nor even a single mathematical symbol. And on the basis of a postgraduate's twenty-seven-page paper, evolutionary biology and contemporary intellectual thought effectively abandoned the previous one hundred-year attempt to apply orthodox natural selection to the human animal, and took us right back to the days of Lord Kelvin and his belief that while humanity evolved, we evolved under uniquely different rules. Perhaps little wonder that John Maynard Smith told the historian of science Ullica Segerstrale that if he talks to a sociobiologist "like Wilson and Trivers for an hour or two, I become wildly hostile" (Segerstrale 2000, p.241).

Turning from Trivers to Richard Alexander, a zoologist and another early proponent of sociobiology, we hear that "population-wide indiscriminate beneficence might also evolve when small 'populations' are regularly composed of relatives related to a similar degree, and if the individuals of other populations are never contacted" (1987, p.100), and that this could be a basis for explaining human large-scale cohesion and generosity. Sociobiologists were positing completely new mechanisms to explain their ideas, such as indirect or third-party reciprocal altruism, where returns may eventually come back from society at large. Hence Alexander: "Moral systems are systems of indirect reciprocity" (p.77). Beyond Wilson, Trivers, Alexander and Lumsden it is often difficult to find a sociobiologist who was a trained biologist, but let us return briefly to Michael Ruse, the philosopher who highlighted Wilson's weakness for group-selectionist theorising. According to Ruse "one will probably function most efficiently when one has no hope of return at all" (1989, p.231). Ruse celebrates the good fortune that "we humans should just so have happened to have evolved to that very morality which is endorsed by God" (p.272). Even for the group-selectionists, chimpanzees live in a world without compassion, and there is no shred of evidence that Japanese macaques have ever gone out of their way to help a severely deformed conspecific in its monumental struggle for existence. And chimpanzees live in such a callous world, and Japanese macaques never lift a finger, for very good, and very efficient, gene-selectionist reasons. Nevertheless for Ruse, Wilson and Trivers human DNA evolved to be the polar opposite of that world without a shred of compassion; we evolved Christian compassion, "that very morality which is endorsed by God". Exactly as the devoutly Christian Lord Kelvin had claimed one hundred years earlier, and Darwin and Huxley had denied was evolutionarily possible.

Sociobiology fell into disfavour by the mid 1980s, though largely for the wrong reasons. As the award-winning science writer and sociobiologist Robert Wright puts it, the sociobiology brand had become toxic – he uses the term "tainted" – by the mid 1980s. Wright says that Wilson's sociobiology "went underground" and emerged as evolutionary psychology in the early 1990s; "most practitioners of the field he defined now prefer to avoid his label" (1994, pp.6–7). But the problem was that evolutionary psychology inherited the core conviction of human sociobiology, the same distorted and confused explicit and implicit group-selectionism. Morality was still argued as being a biological adaptation, albeit often a unique adaptational by-product. So here is Wright's explanation, in his work *The Moral Animal: The New Science of Evolutionary Psychology* (and note we are here "the" moral animal, because for the psychologists and assorted social scientists who had now taken over sociobiology, human evolution was once more and uniquely tearing up the billion-year rulebook). "But the evolution of sacrifice may have grown more complex with time and fostered a sense of group obligation" (1994, p.207), and "reciprocal altruism has extended the sense of obligation – selectively – beyond the circle of kin" (p.212). "In sum, the best guess about valor in wartime is that it is the product of mental organs that once served to maximize inclusive fitness and may no longer do so" (p.391).

Other primates may live in that world without compassion, but gene– and individual-selectionist evolution was again being argued to have left humans as the singularly compassionate and moral animal. Sociobiology, at least when explaining human morality and human group sizes, was a form of confused group-selectionism while still claiming to be built on the gene-selectionist work of Williams and Maynard Smith. Hence Wright calls George Williams "perhaps the closest thing there is to a single founding father of the new paradigm" (p.151), while Harvard's Steven Pinker calls George's *Adaptation and Natural Selection* "the founding document of evolutionary psychology" (1997a, p.56). As Williams's *Nature* obituary put it, "his major contribution, the theory of gene-level natural selection, left a profound and enduring stamp on fields from sociobiology and evolutionary psychology to behavioural ecology" (Meyer 2010, p.790), and even if some of those fields never really attempted to understand his or Maynard Smith's mathematics.

The psychologists' updated version of sociobiology – or as *Scientific American*'s John Horgan put it, the updated version offered by "psychologists, anthropologists, economists, historians and others" (1995, p.151) – is generally seen as having been ushered in by the anthropologist Donald Symons' 1979 book *The Evolution of Human Sexuality*, and then more forcefully by Jerome Barkow, John Tooby and Leda Cosmides' 1992 edited compilation *The Adapted Mind*, the latter becoming the bible of the field. But it was Symons' contribution to the edited work, "On the Use and Misuse of Darwinism in the Study of Human Behavior", that really showed this son of sociobiology to be old wine in new bottles. "In fact, since the adaptations that underpin human behavior were designed by selection to function in specific environments, there is a principled Darwinian argument for assuming that behavior in evolutionarily novel environments will often be *mal*adaptive", Symons wrote, with the emphasis his (p.154). Or, to translate, the adaptation that produced behaviour A 100,000 years ago might actually manifest itself as one hundred and eighty degree contrary behaviours X, Y or even Z, once our peripatetic ancestors left the savannahs and the prairies.

Robert Wright accuses Darwin of being "misguided", and of making "his big mistake", when he analysed the human animal (1994, p.183), but let us return to the LSE's Helena Cronin who if you remember told us earlier that Darwin "lets us down" on the human animal. "But for some of our genes", writes Cronin, "our modern environment is likely to metamorphose their phenotypic expression from that which natural selection originally smiled upon. And genes for behaviour are the most prominent among these. An animal that is adapted to dwell nomadically in smallish bands ... much of that animal's behaviour, however, is likely to change beyond recognition" (1991, p.329). Completely beyond recognition, apparently. Genes for cannibalism and the inability to live in groups larger than about one hundred were now being phenotypically expressed as veganism and fanatical allegiance to vast religions. "There is no need to assume", she writes, and as with Darwin, "that we have to depend on cultural evolution if we are to rise above the selfishness of our genes" (p.369). Actually there is, unless you are among those who, to recall Gould and Kropotkin, "do wish to find a basis for morality in nature and evolution".

Cronin claimed on BBC Radio 4's *Analysis* series in May 1997, emphasis being hers: "reciprocal altruism can be the font of *vast* altruism, self-sacrifice and *genuinely* societal values". Harvard's Steven Pinker is another well-connected cheerleader for the son of sociobiology, but for Pinker it is Cronin who is dead wrong to suggest reciprocal altruism can be the font of vast altruism, because he claims it is really Hamiltonian kin-directed selection that is the font of vast altruism, and even though Hamilton said it wasn't. There is a "cognitive twist" (Pinker, 2012) because humans need "environmental cues", which can include that we come to falsely view complete strangers as kin, and hence Hamilton's inclusive fitness model continues to apply to these "fictive ... faux-families". "Thus people are also altruistic toward their adoptive relatives", which, remember, we have somehow mistaken for our kin. But it is nonsense to argue that falsely viewing fictive faux-families as family through cognitive mistakes is somehow an adaptational by-product, not just because "the altruistic adoption 'strategy' is not an evolutionarily stable strategy" (Dawkins 1989, p.103), but also because "a long memory and a capacity for individual recognition are well developed in man" (p.187). Yet we hear from Pinker that these "extended" families and "illusions of kinship" can apply to fatherlands, fraternities and occupational brotherhoods.

But there was to be a final twist in the saga of the implicit group-selectionism that had marked sociobiology from the earliest days. In 2007 Ed Wilson stunned many by abandoning his supposed earlier support for Hamiltonian kin selection as the key answer to natural world altruism when he subscribed to David Sloan Wilson's model of trait-group selection for the evolution of eusociality, or high-level social organisation, in both social insects and man. Everyone seemed to conveniently misremember the fact that the group-selectionism in Edward O. Wilson's original 1970s work was not just implicit, it was often wholly explicit. It was in no one's interest to recall that both Gould's multilevel-selectionist tradition and Dawkins' gene-selectionist tradition had largely turned a blind eye to Wilson's original group-selectionist writings, even if the two schools had very different reasons to misremember. In 2010 Ed Wilson and two co-authors (see Nowak, Tarnita and Wilson) went further to hammer home their argument that Hamilton's work on inclusive fitness was insufficient and that eusociality lay in multilevel selection and mutations that prescribe the persistence of the group. Patrick Abbot and more than one hundred biologists wrote a reply in *Nature*, claiming "a misunderstanding of evolutionary theory and a misrepresentation of the empirical literature" (Abbot *et al.*, 2011). The attacks continued, or as the *Guardian* put it: "biological warfare flares up again between EO Wilson and Richard Dawkins" (Johnston, 2014), after Wilson described Dawkins as a science journalist. Wilson claimed he had now "abandoned" the gene-selectionism of Williams and Maynard Smith, even though neither he nor any other sociobiologist had ever actually been within this tradition.

Human explanation:	Requires a change to the template of evolution?	Fails the parsimony test?	Invokes new evolutionary processes?
Darwin / Genic selection	NO	NO	NO
Kropotkin / Group selection	YES	YES	NO
Kelvin / Sociobiology	YES	YES	YES

Yet just consider the bigger picture of what was going on here. The academic generally seen as the father of sociobiology, or at least the father of modern-period sociobiology, the biologist who had tried harder than anyone to find a mathematical basis to human sociobiology, was finally admitting that sociobiology had never been offering a possibility for the evolution of human morality, or an explanation for human group sizes. Was suggesting that genetic group-selectionism, a model that few biologists subscribe to any longer, seemed to offer the only theoretical model for the evolution of morality at the genetic level. The only chance for the evolution of morality in humankind at the genetic level, and thus by implication the only chance for the evolution in extraterrestrial life of morality at a biological level. So effectively it is not just Darwin and the gene-selectionists saying that under only a single inheritance mechanism there appears to be no workable model for the evolution of morality in extraterrestrial intelligent life, albeit we will have to consider this point in more detail.

Having filled in the theoretical background – and hopefully started to explain to you why so many important implications have gone under the radar for the last few decades – it is finally time to consider what is at all possible with a single inheritance mechanism. Where we shall discover that while Darwin realised that we must probably wait for a second inheritance mechanism if we wish to meet real extraterrestrial intelligence, the group-selectionists and the sociobiologists in contrast will conclude that we are highly likely to meet extraterrestrial intelligences operating under only a single inheritance mechanism. It is just that for the group-selectionists and the sociobiologists, anywhere between 90% and 99% of those extraterrestrial intelligences will be utterly homicidal.

3
E.T. TYPE I – A SINGLE INHERITANCE MECHANISM

> "The universe we observe has precisely the properties we should expect if there is, at bottom, no design, no purpose, no evil and no good, nothing but blind, pitiless indifference. … DNA neither knows nor cares. DNA just is."
>
> – **Richard Dawkins**, *River Out of Eden* (p.133)

So far, Darwinism has given us cosmic dinosaurs, given us a galaxy teeming with complex plant and animal life, and whatever interpretation of Darwinism we subscribe to. But now we comprehend the three competing evolutionary interpretations it is time to turn to intelligent life, to deliberative life. "When we talk about aliens, we talk about the search for extraterrestrial life; we kind of mean E.T., don't we? We mean something that we can talk to", says Brian Cox. So, logically, what can orthodox Darwinism tell us about E.T.? And what can the group-selectionists and sociobiologists separately offer us in the way of E.T.?

We need to return to the problem of altruism, which also happens to be intimately linked to the problem of group size and group cohesion, which itself then feeds directly into the problem of intelligence. And whether selection acts at the level of the gene or the individual, altruistic sacrifice would at first glance appear to cause a problem for Darwin and Wallace's theory of evolution by natural selection. Nevertheless Darwin realised that there were forms of apparently altruistic and even self-sacrificial behaviour that could pay dividends to the individual so behaving. In the orthodox gene– and individual-selectionist tradition represented by Darwin, Williams, Maynard Smith and Hamilton, natural world altruism has been explained through two main mechanisms known as kin selection and reciprocal altruism.

Kin selection explains the evolution of altruistic characteristics towards close relatives, as a gene for altruism can spread because it enhances its own replication through its survival effects on the relatives. The understanding that animals sacrifice for immediate kin was of course central to Darwin's theory, with fitness defined in terms of simple biological success. But in 1964 kin selective altruism would become one of the building blocks of modern genic selection theory after the publication of W.D. Hamilton's work. William, or Bill, Hamilton suggested that animals seek to maximise not their own fitness, but rather their own "inclusive" fitness. The understanding is that genes pass into future generations not only through direct offspring, but also through relations. In normal diploid organisms like primates, a parent shares a genetic relationship of one half with its child. But the parent also shares such a relationship with its siblings because of inheritance from their shared parents. It will statistically share a quarter relationship with the sibling's offspring, and so on. Relations, and not just offspring, carry an individual's genes into subsequent generations, and so it makes sense, from the "point of view" of an animal's genes, for a degree of sacrifice for near relatives.

Hamilton took these relationships – which had earlier been noted by John Maynard Smith's teacher the geneticist J.B.S. Haldane in a 1955 paper where he wrote of sacrifice that "if you save a grandchild or nephew the advantage is only two and a half to one. If you only save a first cousin, the effect is very slight" (p.44) – and provided a rigorous mathematical formulation of the altruism that could, and could not, be expected under such genetic inheritance. Kin-directed altruistic behaviour can be maintained under selection pressures provided the cost of that behaviour to the altruist (in terms of reduced personal fitness) is less than the benefit of the behaviour to kin (in terms of inclusive fitness) multiplied by the coefficient of relatedness. Hamilton further extended the work to encompass the haplodiploid sex inheritance system, and provided a mathematical rationale for social insect eusociality in terms of the even closer three-quarters average genetic relationship between sisters. Darwin himself had considered issues like neuter insects, and the problem of insect sterility, in the earliest editions of *Origin*, where he wrote, "this difficulty, though appearing insuperable, is lessened, or, as I believe, disappears, when it is remembered that selection may be applied to the family, as well as to the individual, and may thus gain the desired end" (1859, p.258).

Reciprocal altruism is the exchanging of altruistic favours such that each benefits more from co-operating than it would from not co-operating, or "you scratch my back, I'll scratch yours". Darwin wrote that "social animals perform many little services for each other: horses nibble, and cows lick each other, on any spot which itches: monkeys search for each other's external parasites" (1871, Pt. i, pp.74–5). Animals also render more important services, he continued: pelicans fish in concert, wolves hunt in packs, and "social animals mutually defend each other". Animals do indulge in co-operative acts towards those that may not be closely related, not through Kropotkin's benevolent instinct but because the acts are for the self-interested benefit of each party, and Williams referred to this in his earliest work. "A consistent interaction pattern between hens in a barnyard is adequately explained without postulating emotional bonds between individuals. One hen reacts to another on the basis of the social releasers that are displayed, and if individual recognition is operative, it merely adjusts the behavior towards another individual according to the immediate results of past interactions" (1966, p.95). So this is just Darwin's trading of favours. Favours can be returned immediately, or at a later time provided the animals can remember interactions and recognise parties, as hypothesised by Darwin, and now termed "delayed" reciprocal altruism. Such mutually beneficial actions therefore further an individual's genes, and a mutation that first promotes such behaviour can be positively selected.

Examples of reciprocal altruism in the natural world are numerous, and include the mutual grooming of primates, and the oft-cited actions of the cleaner fish. Various species of cleaner fish can occupy specific locations, and offer services to often larger fish and other marine creatures by cleaning them of parasites and unwanted particles. The cleaner fish thereby obtains a reliable food supply from the visitor, as well as a degree of protection from the larger fish. The cleaned fish benefit from having dead and infected tissue removed. Darwin's performance of "many little services" and mutual defence was then to be greatly developed and given a more mathematical basis through the work of John Maynard Smith, who was able to formulate algorithmic strategies of co-operation and confrontation.

Both kin selection and reciprocal altruism are sometimes referred to as "technical" altruism, to make the point that such altruism is ultimately self-serving. All altruism on the non-human Earth reduces to selfish genetic prudence. So where does this leave E.T., and intelligence away from our planet?

"HE BEGAN TO EAT FLESH FROM THE THIGHS OF THE INFANT"

According to Darwin, when you answer to only a single inheritance mechanism, there are limited behaviours evolutionarily possible. And all the gene– and individual-selectionist models tell us that we never out-evolved – could never have out-evolved – Humphrey's billion-year genetic pattern.

> "The dominant male, Humphrey, held a struggling infant about 1.5 yr old, which I did not recognize. Its nose was bleeding, as though from a blow, and Humphrey, holding the infant's legs, intermittently beat its head against a branch. After 3 min, he began to eat flesh from the thighs of the infant, which then stopped struggling and calling."
>
> – **David Bygott**, "Cannibalism among wild chimpanzees", *Nature* (1972, p.410)

In his book *Plan & Purpose in Nature*, George Williams notes that it was the anthropologist Sarah Blaffer Hrdy who was the pioneer in bringing the prevalence of monkey infanticide to the attention of both biologists and the general public. Hrdy found infanticide to be the "single greatest source" of the up to 83% infant mortality rate among the langur monkeys she studied at Abu in India, and her 1977 article, "Infanticide as a Primate Reproductive Strategy", was greeted, says Williams, "with outraged disbelief by many readers who refused to believe that adult males' attacks on infants could be adaptive and normal" (1996, p.218).

Hrdy's groundbreaking paper on monkey infanticide argued that infanticide by males could no longer be dismissed as abnormal. Her own observations on langur monkeys, allied to the work of others including David Bygott, "led me to reject my initial crowding hypothesis in favor of the theory that infanticide is adaptive behavior, extremely advantageous for the males who succeed at it" (1977, p.43). Infanticide was the single dominating source of the very high infant mortality rate, but equally shocking to her were the instances of mothers abandoning their butchered infants soon after or even before death. In the paper Hrdy considered whether, as some had suggested, this was caused by a mother's fear, but rejected that answer in favour of the more hard-nosed gene-selectionist understanding around since the late 1960s: "It is far more likely, however, that desertion reflects a practical evaluation of what *this* infant's chances are, weighed against the probability that her next infant will survive" (1977a, p.286).

While cannibalism subsequent to infanticide is still rare in primates, ape researcher Mariko Hiraiwa-Hasegawa notes that the frequency of cannibalism in chimpanzees is now known to be "exceptionally high" (1992, p.329), and that a female chimpanzee and her adolescent daughter initiated three separate acts of cannibalism as witnessed by Jane Goodall and colleagues in Gombe National Park between 1975 and 1976. As Hrdy put it in 1977: "we are discovering that the gentle souls we claim as our near relatives in the animal world are by and large an extraordinarily murderous lot" (p.46). Sarah Hrdy's paper also cited others' recent findings, from Dian Fossey's observations of infanticide in African gorillas to Bygott's graphic recounting of cannibalism in Tanzanian chimps that we saw above. "This female and her infant were immediately and intensely attacked by the males. For a few moments, the screaming mass of chimps disappeared from Bygott's view" (p.47). Hrdy recorded that when Bygott managed to relocate them, the strange female had disappeared, before one male reappeared holding the struggling infant. In contrast with normal chimp predation, this assaulted and butchered infant, and cannibalised corpse, was nibbled by several males but never actually consumed.

Bygott's paper in *Nature* in 1972 was one of the very first to catalogue chimpanzee cannibalism, and it detailed also the levels of extreme violence against females, and how even low-ranking group members indulged in such behaviour. "Without apparent warning", the older of the two females was viciously "attacked by the five adult males in my group" (1972, p.410). Times have changed in biology, partly due to better field research, but also through a fuller appreciation of the sorts of behaviour that selection acting at the level of the gene can be expected to code for. DNA neither knows nor cares. Only a decade after Bygott was writing, the geneticist Steve Jones would also report in *Nature* that cannibalism had by then been recorded in more than 1,300 species of animal and was often the primary cause of mortality: "there has grown up in biology the comforting supposition that nature is not really red in tooth and claw" (1982, p.202). As George Williams concluded in his own review with James Paradis, "simple cannibalism is the commonest form of killing, and Polis's 1981 review indicates that it can be expected in all animals except strict vegetarians" (1989, p.202). "The story of the forest or coral reef is a tale of relentless arms races, misery, and slaughter", writes Williams (1996, p.214), and he noted that other wild animal populations are many thousands of times more likely to kill than are humans from even the most murderous of American cities. Stephen Jay Gould had the same point to make in his essay "Ten Thousand Acts of Kindness". Ethologists, Gould noted, describe organisms as peaceful if tens of hours go past with only one or two aggressive encounters. Consider the many millions of hours we can log for most people on most days with nothing more than a raised middle finger every once in a while, he wrote. "*Homo sapiens* is a remarkably genial species" (reprinted in 1993, p.281).

"Mountains of data on parasitism and predation (including cannibalism) in nature could be amassed to document the enormity of the pain and mayhem that arise from adaptations produced by natural selection", says Williams (1996, p.216). In his academic work Williams has been more detailed, and in one paper he devoted over seven pages simply to documenting the data already collected by field researchers. "Besides adultery and rape, just about every other kind of sexual behavior that has been regarded as sinful can be found abundantly in nature. Brother-sister matings are the rule in many species (Hamilton 1967)" (1988, p.395). Gone is the bright optimism of Darwin's day where some could still argue that nature might teach moral lessons to humankind. "With what other than condemnation is a person with any moral sense supposed to respond to a system in which the ultimate purpose in life is to be better than your neighbor at getting genes into future generations, ... in which that message is always 'exploit your environment, including your friends and relatives, so as to maximize our (genes') success', in which the closest thing to a golden rule is 'don't cheat, unless it is likely to provide a net benefit'?" (1996, pp.213–4).

Williams catalogued the differential infanticide strategies nature has developed in ground squirrels. "A male may raid a nest to kill and eat one of the young. A female may raid the nest of a competitor and kill all the young (but not eat them)" (Paradis & Williams 1989, p.202). While it is always more difficult for a male to keep track of which offspring he has fathered, even usually in species where adults pair for life, it is much simpler for the female to identify the threat posed to the hegemony of her genes and do something about it. And rape was a strategy that was widely expected by selfish gene theorists even before the observations began flooding in. In one of the earliest detailed papers on avian rape, Pierre Mineau and Fred Cooke noted in 1979 that even the briefest absences by her mate could make the female snow goose vulnerable to rape by neighbouring males. Rape makes mathematical "sense" in an animal kingdom when the only thing that drives life's algorithm is the need to get your sperm to fertilise as many eggs as possible. Nature has made the male cautious about leaving his mate unguarded for long, because in addition to rape he runs the risk that she will seek a finer gene pool than is offered by his seed. Two can play the game of enhancing gene survival.

Such behaviours are not by-products of naturally selected behaviour; they are the dynamics of natural selection. An animal that is busy protecting its mate from rape this season will be busy raping others' mates next season. As Mineau and Cooke noted of their snow geese, personal status, familiarity and coupling matter little in the game of gene preservation. This is the universal genetic logic of natural selection where only opportunity matters. "The rapist also was usually a known territory holder (84 percent), commonly a neighbour. Rapists seem to capitalize on attendant male absenteeism. ... An absent male is himself usually (73 percent) raping another female or witnessing another rape. Indeed, 'gang rapes' usually occur when spectators at a rape attempt use the disturbance to join in the melee. Up to 80 spectators have been seen at a rape attempt" (1979, pp.282–3). The authors observed nine rape attempts in two days upon one female left unattended by her mate, while Williams adds that "an unguarded female mallard may be attacked so persistently by gangs of males that she drowns" (1988, p.394).

Cannibalism is as common in marine vertebrates as it is in birds and dry-land vertebrates. Almost 200 years ago geologists were finding evidence of cannibalism in the petrified faeces of 200 million-year-old marine reptiles (Buckland, 1835), and today it is marine cannibalism that provides some of the most stunning examples of the benefits of cannibalism exceeding the costs. In 1948 intrauterine, or within-the-womb, cannibalism was discovered in the sand tiger shark, *Carcharias taurus*, also known as the grey nurse shark, where the largest and strongest embryos eat their weaker siblings in the process of adelphophagy (literally *consuming one's brother*). Sand tigers have two oviducts, and the first young in each oviduct to reach 6 cm swims to the uterus where it feeds on its siblings; a large part of the reason that it is commonly around 100 cm at birth. The major benefits from an evolutionary point of view are that surviving sand tiger young experience very rapid growth, are very active predators at birth, and often have a considerable size advantage over other predators. The pike-like freshwater game-fish the walleye has provided examples of cannibals within cannibals within cannibals, as larger walleyes were found to have eaten smaller walleyes, which had eaten still smaller walleyes, for at least a four-fold cycle. Or there are Atlantic bluefin tuna, where more than 90% of the tuna larvae analysed had at least one cannibalised prey in their stomachs (Uriarte *et al.*, 2019). Cannibalism in marine mammals is now well documented. Cannibalism has been seen among grey seals in Canada and elephant seals in Argentina, but even though sea lions have been studied since at least de Bougainville's voyages in the 1760s and Johann Forster's voyages in the 1770s – the latter's studies mentioned by Darwin in *Descent of Man* – it was only as recently as 1999 that cannibalism was first recorded in sea lions, when twenty-four cases were seen in just twelve weeks in one colony of New Zealand sea lions.

All of the above makes perfect sense when selection is at the level of the individual, or at the level of the gene. Richard Dawkins tells us about blackheaded gulls within five pages of beginning *The Selfish Gene*, and as a paradigm of gene-selectionist behaviour. "It is quite common for a gull to wait until a neighbour's back is turned, perhaps while it is away fishing, and then pounce on one of the neighbour's chicks and swallow it whole. It thereby obtains a good nutritious meal, without having to go to the trouble of catching a fish, and without having to leave its own nest unprotected" (1989, p.5). Natural selection is about sacrifice for those that carry your genes, not what we would understand as love. A mother may fight savagely to protect her offspring, but when she knows the game is lost her behaviour changes immediately. Weak offspring are often killed if this is the most efficient strategy. And no sustained grief or bitterness, since one gene vehicle is likely to be as good as another, and just as much of a biological tool. A mother chimpanzee will mate contentedly with the killers of her infant. Smaller mammals, with less access to reliable sources of food, and worn down by a long period of having to suckle, will kill and eat their own offspring rather than allow them to fall into a predator's hands. This is because a healthy mother will get a chance to mate again next season, so it is time to build herself up rather than waste energy fighting the inevitable. Why risk serious harm to yourself when the product is replaceable? In nature, infants are important, but you can have others next season, so they are only worth fighting for up to a point. The point when the costs outweigh the benefits. "There is no charity in nature", writes Steve Jones, and "the laws of the animal world are ruthless. Plenty of parents kill their children, and plenty of children murder their sibs. … The surplus young are a biological insurance policy, and when times look good their parents begrudge the cost of cover" (1999, pp.160, 172).

NO EXCEPTIONS TO THIS RULE ...

At this stage many will be looking for an escape from relentless individual-selectionist logic. What about dogs? they will ask. Loyal, self-sacrificing, obedient; surely man's best friend breaks this depressing and unrelenting pattern? We have certainly managed something a bit special with dogs. Artificial selection to retain juvenile characteristics has combined with our ability to take our place as their pack leader. What has been termed dogs' hypersociability, or exaggerated gregariousness, has been linked to molecular mechanisms for "the extension of juvenile behaviors into adulthood" and enhanced bonding (vonHoldt *et al.*, 2017). But domestication, like civilisation, is a thin veneer, and even untold generations of intense artificial selection in the descendants of the wolf have been barely able to mask the amorality of the natural world. As Steve Jones notes in *Almost Like A Whale:* The Origin of Species *Updated*, "by owning a dog, any dog, men welcome into the home a beast that preserves much of its primordial self. Overgrown juveniles though they are, evolution by human choice has not removed the instincts of their ancestors. ... Like wolves, dogs attack the weak, be they children, old, or drunk". Packs of feral animals have pulled infants from bicycles and eaten them, he writes, "and a mere half-dozen beagles, dachshunds and terriers once devoured an eighty-year-old woman. The homicidal packs relive their past" (1999, p.41).

> "July 24, 1990: *Ntologi* had in his hand the 5-month-old infant of *Betty*, a primiparous immigrant. The infant was still alive. *Ntologi* began to bite on the fingers of its right hand. He struck the infant against a tree trunk, and also dragged it on the ground as he displayed. As a result the infant was finally killed. After a while, *Musa* began to feed on the fingers of the left hand, and *Bakali* and *Lukaja* fed on the toes. ... In total, ten adult females and eight adult males came to eat."
>
> – **Hamai et al.**, "New records of within-group infanticide and cannibalism in wild chimpanzees", *Primates* (1992, p.153)

A couple of decades ago bonobos, once called pygmy chimpanzees, were regularly being touted as a paradigm for human biological development, and as an exception to the above great ape behaviour pattern. Bonobos were apes that, as many newspaper articles of the time had it, "make love, not war". As Richard Dawkins wrote in *Unweaving the Rainbow*, group-selectionists often react indignantly to the idea that nature is genetically selfish, himself singling out the primatologist Frans de Waal, author of the tellingly-titled *Good Natured*. De Waal "is distressed at what he mistakenly sees as a neo-Darwinian tendency to emphasise the 'nastiness of our apish past'. Some of those who share his romantic fancy have recently become fond of the pygmy chimpanzee or bonobo as a yet more benign role model" (1998, p.211).

Ethologists studying animal infanticide have also noted that "De Waal tends to dwell on the niceties of animal societies – 'Survival of the kindest' (1998). ... There is still fundamentalistic resistance in the scientific community to the idea that animals act selfishly – particularly when it comes to infanticide" (Sommer 2000, p.13). Volker Sommer was himself writing the introduction – entitled "The Holy Wars About Infanticide. Which Side Are You On? And Why?" – to van Schaik and Janson's edited collection of more recent animal infanticide reports.

De Waal and others had originally suggested that bonobos might be better models for early humans, if one can overlook the fact that bonobos regularly indulge in sex in every possible combination, and have no exclusive sexual orientation. But Craig Stanford used the pages of *Current Anthropology* in 1998 to point out that the number of field observation hours on bonobos was a small fraction of the hours spent field observing chimpanzees, due to bonobo temperament, the impenetrable forests they inhabit, and regional instability. Stanford had also noted that half of all encounters between bonobo communities do still result in aggression of some sort, and that observer bias, including the overemphasis on captive, rather than field, observation, may be misleading the public and scientists alike. While infanticide and lethal intercommunity aggression are both common in chimps, they are still restricted in time. Furthermore, natural selection will tend to come up with a counter-strategy where possible; if males are going to kill unrelated infants, females will tend to have been selected to steer clear of males for the period when infanticide can advance oestrus. Similarly, intercommunity conflicts are predictable but tend to be short-lived. Consequently, chimpanzees had to be observed for more than fifteen years before lethal intercommunity aggression was finally witnessed.

Many of Stanford's points were well taken by other bonobo researchers, especially the field researchers. Katharine Milton reminded her colleagues that: "we do not have to look far into the past to recall how, as more field data emerged, the sunny image of the playful, fruit-eating chimpanzee at Gombe was gradually revealed to have a darker side" (1998, p.412). Takayoshi Kano, head of the bonobo project at Wamba, noted that lethal bonobo intercommunity aggression might indeed one day be witnessed, citing his own observation of "a severe laceration" on one young adult that got separated from his main party for a few days (1998, p.410). And even de Waal was quick to point out (1998) that he himself had said that infanticide may yet be witnessed in bonobos when more study had been undertaken. De Waal recalled that "science has erred before with a range of so-called peaceable species, from gorillas to dolphins. ... Generally, such idealizations mean that something highly significant has been overlooked or, worse, covered up" (1997, p.84). Such hedging of bets seemed wise. Noting that infanticide by males is most advantageous where lactation is long relative to gestation, van Noordwijk and van Schaik (2000) had predicted that all great apes are vulnerable to infanticide even though, while well recorded in gorillas and chimpanzees, infanticide had not then been witnessed in wild orang-utans or bonobos. Frequency of expected attacks, however, is largely a function of the counter-strategies females have evolved. Female orang-utans are semi-solitary. Thus although female orangs are exposed to male attacks – including frequent documented rape – while actively receptive to sexual encounters, after giving birth, mothers with infants rarely associate with conspecifics. Frequency of infanticidal attacks is thus expected to be low but not zero.

Since the turn of the millennium the bonobo's image has started to become much darker. Infanticide has still not yet been confidently witnessed in wild orang-utans or bonobos, and some researchers have argued that orang females form special bonds with protector males to guard against infanticide. Others have suggested that concealed ovulation, deliberate paternity confusion, and behaviour, including ranging, may provide reduced benefits from infanticide, fewer opportunities, or just less chance of observation. But in 2009 David Dellatore and colleagues reported two incidences of female orang-utans eating their recently deceased infants (Dellatore *et al.*, 2009). At the same time, the primatologist Andrew Fowler at the Max Planck Institute for Evolutionary Anthropology in Leipzig noted that it "had been suggested in the past that bonobos might feel more sympathy for victims, which is why they didn't hunt monkeys", but that image had been shattered in 2009 when scientists discovered that bonobos do indeed kill and eat monkeys. And in 2010 Fowler and Gottfried Hohmann reported their witnessing of an example of a bonobo mother consuming, along with most other apes in her group, the body of her recently deceased 2.5-year-old infant. During seven and a half hours of observation six of the nine adult females present consumed meat, as did two of the three adult males. Several adult and juvenile males "played with the skin, running and chasing the possessor but without aggression. After most of the softer tissue had been consumed" the mother, Olga, "carried the remains that consisted mainly of skin with one foot and hand attached" (Fowler and Hohmann 2010, p.511). As the BBC put it, Fowler and Hohmann's observation "does further challenge a widely perceived notion that bonobos are an especially 'peaceful' ape species" (Walker, 2010), and Fowler stressed again how few long-term studies of bonobos there have been. In late 2016 Nahoko Tokuyama of Kyoto University and colleagues reported, after having caught on video, similar bonobo maternal cannibalism (Tokuyama *et al.*, 2017). This occurred at two sites in the Democratic Republic of the Congo, Wamba and Kokolopori, and involved mothers merrily eating their dead offspring and sharing the torn-apart carcasses with others, including juveniles and infants. The mother at Wamba, Hide, a high-status central female, "grabbed the carcass in her right hand and suddenly bit off the head". Within ten minutes "she had consumed the carcass's head and most of its torso, but all arms and legs were intact". Almost ten minutes after that, "Hide was still holding one detached limb" (p.9). At Kokolopori, "eight adult females, one juvenile female, and two infants participated in the cannibalism" (p.10). Watch the video produced by BBC Earth Unplugged. This opens (2016) with a bold printed warning that it contains scenes which some viewers may find distressing, followed by the presenter's similar verbal warning, and is produced from Tokuyama's graphic footage. In the video, Hide is shown using her dead infant's forearm – to which its little hand is still attached – as a combined cocktail stick and toothpick.

Tokuyama and other colleagues (Tokuyama *et al.*, 2021) have in contrast recently encountered bonobo adoption of two infants from across another social group. Adoption of out-group members is very rare in primates, though it should be appreciated that domesticated mammals living with humans have adopted interspecies (cats have adopted puppies, guinea pigs and even ducklings, while dogs have adopted kittens). Intergenus adoption can sporadically happen in the wild, and a traumatised lioness may have adopted more than one orphan antelope, an inexperienced bottlenose dolphin nursed a melon-headed whale calf for three years (Carzon *et al.*, 2019), and a lactating lioness adopted and suckled a leopard cub (Mittal *et al.*, 2020). As Williams once noted to me: "My impression is that animal adoptions do happen occasionally. ... A parent is elaborately programmed to be taking care of an infant, which dies. It seems unreasonable to expect the program to suddenly disappear. Couldn't it briefly lead to some maladaptive behaviours?" (letter, 27 November 2002). An infant marmoset was for over a year adopted by wild capuchin monkeys (Izar *et al.*, 2006), albeit its significantly smaller natural size meant it sometimes struggled to keep up, when its cries ("distinctive vocalizations", p.698) were all but ignored by the rest of the group. The bonobo out-group adoptive behaviour thus has been seen before, in monkeys, albeit there it is more likely a misplaced mothering instinct. This bonobo behaviour is somewhat different, though Tokuyama *et al.* did note that one of the two mothers also had biological offspring and did not devote the same level of maternal care towards her adoptee, grooming her natural offspring "significantly more frequently". Such behaviour could be the result of attempts to strengthen out-group bonds or even strengthen the alliances or fitness of the in-group, because bonobos have somewhat more tolerant inter-group behaviour, with sexually immature females freely emigrating across groups. Chimpanzee orphans seem to be similarly often adopted for strong social relationship reasons, rather than genetics (Hobaiter *et al.*, 2014; Samuni *et al.*, 2019).

But to argue that bonobos are relatively peaceful, and perhaps therefore even proto-moral, is to misunderstand the mechanism of both individual-selectionism and gene-selectionism. While the group-selectionist model occasionally argues differently (and thus may have somewhat distinct implications for extraterrestrial intelligence that we will need to consider later), for both the gene-selectionist and human sociobiologist models "there is no charity", to recall Steve Jones, in non-human nature. Like orang-utans, the reason that bonobos rarely, if ever, commit infanticide is purely because they do not get the opportunity, relative to cannibalism where they do get the opportunity. Bonobo groups are largely female-dominated, whereas chimpanzees are overtly male-dominated. In bonobos, females get to set the rules. One rule they set was sex. De Waal comments that what the female bonobos are doing might be construed as an anti-infanticidal strategy. "They have managed to make paternity so ambiguous that there is little to fear. Bonobo males have no way of knowing which offspring are theirs and which not" (1997, p.122). Male chimps, de Waal notes, have a tendency to kill identified unrelated infants, so it is no wonder that chimp females stay away from large gatherings of their species for years after having given birth. No such worries for female bonobos, however, because a male bonobo who knew his offspring "would have to be a genius"; "if one had to design a social system in which fatherhood remained obscure, one could scarcely do a better job than nature did with bonobo society" (pp.120–1). Furthermore, noted de Waal, since bonobo females tend to be dominant, attacking them or their offspring would be a risky business. Most likely if a male were to be perceived as a threat to any infant, females would band together in pre-emptive defence.

Chimpanzees are already acknowledged by de Waal to live in "a world without compassion", yet bonobos also live in this individual-selectionist and gene-selectionist world without compassion, and bonobo behaviour can be every bit as violent as chimpanzee behaviour. Primatologists have been intrigued by the high rate of physical abnormalities in the male bonobos at Wamba. As Amy Parish told de Waal in interview, this could have something to do with the females. Females in captivity establish dominance over male bonobos by overt aggression. At one zoo, noted Parish, the females occasionally held down the male and attacked him, and had bitten off parts of his fingers and toes. At another zoo the alpha female had a similar relationship with her adult male: "It is assumed that she once bit his penis almost in half", Parish reported (De Waal 1997, p.115). Wakefield *et al.* recorded a range of behaviours when wild bonobos shared antelope meat but this included "frequent aggression" (2019, p.186), also noting that females usually kept control of the carcasses, and in contrast with chimpanzees they generally thwarted theft attempts on the carcasses by males.

At least as early as 1989 George Williams had been predicting that cannibalism would be recorded in all great apes, and indeed that it could be expected in all animals except strict vegetarians.

Due to his illness Williams never became aware that cannibalism had finally been recorded in bonobos, as Fowler and Hohmann's report did not appear until 2010. With cannibalism recorded in chimps and orang-utans, and Dian Fossey having apparently found gorilla remains in the faeces of other gorillas in the 1970s, and notwithstanding that gorillas stick mainly if not strictly to a vegetarian diet, bonobo cannibalism seems to round off the great ape set. Williams had made a *genetic* prediction about cannibalism based on his and Maynard Smith's updating of Darwin's theory, and two decades later field researchers would prove his prediction correct in regard to our closest *genetic* kin.

WHEN SIZE DOES MATTER

Some of John Maynard Smith's most valued work dates from the early 1970s when he and George Price became the first to apply game theory modelling to the natural world. Game theory is a branch of mathematics developed by Morgenstern and von Neumann in the 1920s as a way of formalising social questions and activities, based on the underlying assumption of each player in the game acting rationally. The most famous game theory scenario is called prisoner's dilemma, where co-operation will not evolve in rational self-interested players who never have to repeat the game. But co-operation can evolve when the game is continuously repeated, known as iterated prisoner's dilemma. Under these circumstances long-term expedience is best served by co-operative strategies of enlightened self-interest among players who continually have to interact. And such low-level and calculated co-operation is exactly what we observe in the natural world, and where natural selection has replaced the rationality of the player and in the form of reciprocal altruism and delayed reciprocal altruism.

Evolution is a struggle for survival, and works because it is. As Williams demonstrated in 1966, any adaptation that somehow managed to evolve for the good of the group would suffer from subversion from within. Organisms within groups are still competing against one another, and the organisms that begin to take the benefits without paying the costs will be favoured by evolution. Those organisms that acted for the good of the group, rather than, at all times, being driven by what is good for themselves, would be taken advantage of and would not be the successful ones. Subversion from within has consequences. Maynard Smith made a major breakthrough when he applied game theory to the natural world to develop the concept of the evolutionarily stable strategy, or ESS. These strategies are the possible group behaviours nature has discovered within the limits set by given genetic inheritance and reproduction systems. For example, natural selection has ensured that for non-inbred and diploid – meaning two complete sets of chromosomes, one from each parent – organisms close co-operation is only possible in small groups; for all other apes, it is only possible in groups of around five to not much more than one hundred or so. An ESS is defined as a strategy which, if most members of a population adopt it, cannot be bettered by an alternative strategy. Such a strategy cannot then be surpassed by an aberrant individual; the strategy becomes stable and cannot be subverted from within. No deviating individual will have a higher fitness or potential for reproduction, and the trait thus becomes fixed. It is the paradigm of game-theoretic modelling, where it is Darwinian selection playing the role of the rational policy selector. "The use of optimization is easier to justify in biology than in economics, because natural selection provides a dynamics which will, subject to constraints, cause a population to evolve towards an optimum", as Maynard Smith noted in his concluding remarks in *Evolution of Social Behaviour Patterns in Primates and Man* (1996, p.291). Groups are made up of competing individuals, and natural selection creates limits on group sizes, dependent on the level of co-operation necessary between the individuals (reciprocal altruism), and the degree of relatedness among them (kin selection, the sacrifice one organism makes for close relatives).

Biological theories of altruism work, in diploid, non-inbred, and non-sterile caste species, for small family groups where there is a high degree of genetic relationship, and for small reciprocating groups where the parties regularly interact and can consequently recognise each other and remember the interchanges. Natural world "tit-for-tat" exchanges can only work under such a system. If one chimpanzee grooms another chimpanzee it is because it is looking for something in return from that other chimp. Group size is a function of many factors, including food availability and security. The type and frequency of prey may necessitate a larger group for regular successful kills, but this larger group means the disadvantage of having to share a kill among a larger number. Open areas require larger groups for protection of members against predators; in a wooded area smaller groups afford sufficient protection because individuals can more easily escape in the trees. Nevertheless, one crucial restraining *upper* limiting factor will be given by the degree of interaction necessary between group members. Wildebeest may sweep majestically across the plains in their thousands, but wildebeest don't have to co-ordinate and share a kill and don't therefore have to work together. In fact, Bill Hamilton wrote a famous paper, "Geometry for the Selfish Herd", and as an "antithesis to the view that gregarious behaviour is evolved through benefits to the population or species" (1971a, p.295). Hamilton's paper, the start of what is called selfish herd theory, and now widely accepted at least for 2-dimensional space, highlighted the evidence that the members within a herd are forever cover-seeking, pushing for the safest positions at the expense of others, including pushing to get to the centre of the herd. Behaviour that necessitates many interactions between group members, in contrast, requires strong group cohesion. This means individuals need to know each other well. Nature has established that for primates such cohesion cannot be left to degree of relatedness alone. Grooming is an essential part of primate behaviour because it allows members to form bonds and get into reciprocating relationships.

But group cohesion in primates is limited. Firstly, there is the need for strong reciprocating relationships that require detailed knowledge of your associates; detailed knowledge that can only be gained and maintained by forfeiting the possibility of a larger, more loosely-knit structure. Secondly, grooming takes time. Primates may spend up to 20% of their time grooming for cohesion. They cannot afford more because over 80% of their available time must be spent hunting and working. So there are reasons for group sizes in nature. Group size is effectively kept stable by the positive and negative forces at play. Hence chimpanzees live in fission-fusion societies of rarely above one hundred members, noting that the group in Ngogo has exceptionally reached 200 members, and they forage in much smaller groups. Bonobos tend to live in similar or slightly reduced groups. Even where we leave the apes and look at the less cohesive – and less closely co-operating – monkeys, we find that baboons will usually live only in groups of a couple of hundred. One species of baboon can live in groups of up to 800, though what we always find with larger groups in the primates is that they will be made up of numerous smaller subgroups, with little peaceful interaction occurring between subgroups. And intelligence has little to do with it; the smarter the primate, often the smaller the groups, in part because smarter animals are better at cheating, and smaller groups more effectively limit the opportunity for cheating and the subversion this creates.

So we have at last found our first rule for intelligent extraterrestrial life, for deliberative life, remembering that we have moved beyond just cosmic dinosaurs and cosmic wildebeest

E.T. TYPE I – A SINGLE INHERITANCE MECHANISM

and now, as Brian Cox reminded us, "we mean something that we can talk to". Transition six, the evolution of multicellularity, almost certainly cannot on its own give rise to E.T., to something that we can talk to, at least not for Darwin, though we will turn at the end of the chapter to the deeply homicidal transition six E.T.s that group-selectionism and sociobiology will both predict. Transition six, for Darwin, cannot really give us something that we can talk to, cannot give us a species that can properly dominate its home world, and certainly cannot give us a spacefaring species, because of the small group size problem in diploid species. Chimpanzees for example do have a form of social communication, just basic emotional grunts and hoots, plus more complex gestures and facial expressions, but nothing close to a language, nothing close to talking. Complex language, dominion, technology; all require a large brain of course, but more than that they require very large groups, and a functional division of labour. Complex language and technology require some to be providing for the basics in life so that others can have the time and resources to indulge in creative thinking and experimentation. "Although some groups, such as honeybee colonies, are functionally organized, most animal groupings are not. They are just mobs of self-seeking individuals", writes George Williams in *Plan & Purpose in Nature* (1996, p.80). Complex language, dominion, technology; all require more than just mobs of self-seeking individuals pushing to get to the safest place in the group or herd.

On its own the evolution of multicellularity, transition six, probably cannot give us E.T., cannot give us something that we can talk to, at least according to Darwin. But nature has found ways to go further on group interaction using only a single inheritance mechanism; has found a way to effectively supercharge that single inheritance mechanism. Nature discovered transition seven.

"A BEHAVIOURAL AND REPRODUCTIVE DIVISION OF LABOUR"

Altruism in nature can be explained through the two main mechanisms known as reciprocal altruism and kin selection, both ideas in different ways touched upon by Darwin. But it was Bill Hamilton who first firmly drew the biological community's attention to inclusive fitness and the importance of kin selection in a seminal two-part paper in 1964. "Hamilton's Rule", as developed in that paper, was applicable to all species by virtue of their relatedness, but the paper is partly remembered for the way it was applied to the genetical asymmetry of haplodiploid Hymenoptera, which includes ants, bees and wasps, but not termites.

Mammals, being diploid, have a double set of chromosomes, one from each parent. In haplodiploid Hymenoptera, diploid females develop from fertilised eggs and have a double set of chromosomes, one from each parent, while haploid males develop from unfertilised eggs and have only a single set of chromosomes from their mother to pass on. Sperm from a male are thus genetically identical. The coefficient of relatedness of mother to daughter has the normal value of 0.5. But the average relatedness between daughters from this male is 0.75, closer than it would be to any offspring they might conceivably have. Helping your siblings – who may number in the hundreds of thousands – can become of overriding genetic importance. Hamilton noted that family relationships in Hymenoptera are potentially very favourable to the evolution of reproductive altruism.

Another route very favourable to the evolution of reproductive altruism is inbreeding because relatedness can rise above the value of 0.5 that normally applies under outbreeding. Thus, noted Hamilton, an individual should be more altruistic than usual towards its immediate kin. Termites, the Isoptera, for example, are not haplodiploid, but can indulge in cycles of intense inbreeding within colonies and outbreeding to found new colonies. Some termites, however, found new colonies with daughters from a queen produced by parthenogenesis; multiple identical clones, effectively, but thereby permitting a far greater number of offspring while avoiding the problems of inbreeding. There are subtle problems to seeking the origin of social insect eusociality in simplistic application of coefficients of relatedness, and the question of insect eusociality is therefore by no means settled. Nevertheless, modern biology generally seeks its answers to the origin and maintenance of extreme insect sociality within (and blending) a limited range of possibilities, from very close genetic relatedness to pheromonal suppression of fertility, parasitic influence, resource dependency, and other ecological factors. Vast offspring production plus irreversible reproductive specialisation, morphologically delineated castes (such as large reproductives, midsize soldiers with powerful mandibles, and small to median workers across several castes), and a high level of hard-wired behaviour may also enable social insects to achieve group sizes impossible within the non-human mammals.

Important to this last understanding is the naked mole-rat. The naked mole-rat is "arguably the closest that a mammal comes to behaving like social insects such as bees and termites, with large colonies and a behavioural and reproductive division of labour", write Bennett and Faulkes in *African Mole-Rats: Ecology and Eusociality* (2000, p.i). The Damaraland mole-rat is the other eusocial mole-rat, with non-reproductive offspring, yet colonies are tiny, with an average size of only twelve members, and a maximum size so far found of around forty members. Damaraland mole-rats have a strong inbreeding avoidance mechanism and have a relatedness coefficient of no higher than the normal 0.5 found in outbred first-degree relatives. In contrast, even more extreme eusociality is displayed by the naked mole-rat, where average colony size is around eighty, and colonies of up to 300 animals have been discovered. Intense inbreeding among naked mole-rats has led to average intra-colony relatedness of 0.81, the highest recorded for a natural mammalian population, and naked mole-rats protect against some of the problems of inbreeding through the discovery of an outbreeding disperser morph or form. Comparison between the two species and other mole-rat species has tempered the initial view that it is inbreeding that explains all in naked mole-rat eusociality (there is one breeding queen, one to three breeding males, and the rest sterile or reproductively suppressed workers). Though levels of relatedness are always an important factor, ecological factors are also thought to be critical. Implicated here is a harsh and unpredictable climate and location producing benefits from assisted living, plus the very high costs constraining dispersal, both in terms of creating a new subterranean environment from scratch and the low probability of finding alternative food and mates.

Species of social insects have colonies that can range from a few hundred individuals even up to several million, so we have finally managed one route to achieve the large group sizes and division of labour which we will need to see, at least according to Darwin, if we are to hope to come across another species that we can talk to, or a species that uses technology, or even a spacefaring species. But there are some points we must consider here. The first is that we have not yet escaped this world of "blind, pitiless indifference", this world of "DNA neither knows nor cares". We are still operating within the constraints of a single inheritance mechanism, and the logic is effectively unchanged. Penelope Kukuk writes about social bees in Elgar and Crespi's survey of cannibalism across diverse taxa. She notes of stingless bees, "these bees are obligately social, and neither queens nor workers can survive and reproduce alone. They are committed to group life with a well defined reproductive division of labour" (1992, p.225).

But although Kukuk touches on queen-worker conflict theory it is left to Crespi himself to provide the main chapter, explaining that colonial living is still all about dominance and crushing dissent, and a bubbling level of subversion. "The queen is selected to maintain her status by suppressing any reproductive tendencies in workers"; "the social situation in colonies changes dramatically when the queen's pheromonally-mediated and behavioural dominance begins to fail in late summer. ... Older workers are usually most aggressive, cannibalistic, and fecund. Fighting occurs between queens and workers, and among workers" (1992, pp.186, 189). And returning to mole-rats, Bennett and Faulkes tell us that there is intense conflict during contests for succession. They call colonies "a reproductive dictatorship", where the breeding female naked mole-rat is typically the most dominant animal in the colony, "and rival challengers for breeding status are usually attacked and killed by the queen" (2000, pp.175–6), who is, remember, often very closely related to those she kills. Naked mole-rats "are very xenophobic to non-colony members" (p.185).

These may not be the "mobs of self-seeking individuals" we saw in the previous sections, but they are still subject to both the extreme violence and the cheating we saw across the diploid world. Maynard Smith and Szathmáry introduce the work of behavioural scientists who have identified the main factors affecting reproductive "skew" within colonies across what has been called the eusociality continuum. The higher the skew, the more some individuals contribute genes to the offspring than others. Three of the key factors to reproductive skew are degree of relatedness, how successful a subordinate would be if it left the colony to go it alone, and "what is the chance that a subordinate would win a lethal fight with the dominant without being severely injured" (1999, p.128), noting that worker castes tend to be "small relative to the queen". Across the continuum it is always a trade-off, with a simmering level of subversion, affected by factors including ecological constraints. In the dwarf mongoose, for example, "dominant individuals do not completely suppress reproduction by the subordinates", and because older subordinates have a relatively high likely success if going it alone, so suppression must be more measured to avoid triggering their exit from the colony. So this is still Darwin's individual-selectionist (and the gene-selectionist) universe of suppression, violence, subordination and grudging trade-offs, not equality, kindness and virtue. As Bernard Crespi notes of eusocial colonies, "some degree of overt conflict is always expected, less 'cheating' arise and begin to spread" (1992, p.200). Subversion from within must always be headed off at the pass.

So with transition seven we do now have the potential for vast interacting groups, and vast cohesive groups are necessary – although not necessarily sufficient – for the possibility of complex communication, planetary dominion, and maybe even a spacefaring civilisation. We may now have our first form of E.T., our first form of intelligent, fully communicating, cosmic life, but we still need to investigate further. Eusocial colonies, with perhaps a bit of additional intelligence, have the potential for both the violence and co-ordination necessary to dominate their lands. Richard Dawkins introduces us to driver ants and army ants, some living in colonies up to 20 million, and all "wandering in enormous pillaging armies. ... Army ants and driver ants, or rather their colonies taken together as amoeba-like units, are both ruthless and terrible predators of their respective jungles" (1986a, p.107). Both cut to pieces anything animal in their path, Dawkins writes, and both have acquired "a mystique of terror". Dawkins notes that as a child living in Africa he was more frightened of driver ants than he was of lions and crocodiles. Or as the Pulitzer Prize-winning entomologist Bert Hölldobler has written, "the foreign policy aim of ants can be summed up as follows: restless aggression, territorial conquest, and genocidal annihilation of neighboring colonies whenever possible" (writing with Ed Wilson 1994, p.59). But although vast, driven, terrible, and even genocidal, to what extent are these more than just a single creature? Dawkins mentioned their amoeba-like units, and also quotes Ed Wilson – "the driver ant colony is an 'animal' weighing in excess of 20 kg" – who was an expert on social insect behaviour before resurrecting modern period sociobiology. "Those gaping soldiers were prepared to die for the queen … simply because their brains and their jaws were built by genes stamped from the master die carried in the queen herself" (Dawkins 1986a, p.108).

Colonies are still subject to a level of subversion but are often referred to as "superorganisms" because of the extraordinary co-ordination and co-operation they can demonstrate. Individuals even actively sacrificing for the good of the colony, Williams' functional organisation mentioned above for honeybees; a functional unity. The characteristic of such colonies is they are not just the mobs of self-seeking individuals, but the larger the colony, the more hard-wiring required, the more we see behaviour "stamped from the master die". The greater the individual intelligence, generally, the smaller the colony. Naked mole-rat colonies are orders of magnitude smaller than some social insect colonies can approach. And remember that colonial living has often evolved where it assists in extreme environments, as with naked mole-rats, limiting the opportunities for much more than simple survival. Everything about colonial living still seems to militate against the creativity and the individualism, the intellectual division of labour, that seems to be necessary for true intelligence and communication, and seems necessary for the development of engineering, technology and science. We have larger group sizes now, but not overly large without hard-wiring, and we still seem to lack much in the way of intelligence, knowledge or creativity.

BENEVOLENT EXTRATERRESTRIALS, AND THE PROBLEM OF HUMPHREY

We now know enough about evolution – whichever tradition one holds to; gene-selectionist, group-selectionist, or sociobiologist – to understand the stark and very real problem we must overcome if we ever hope to encounter benevolent extraterrestrials. The problem of Humphrey's cannibalism. The problem of Ntologi, and Lukaja, and Bakali, and the other dozen or so chimps who came to feed on Betty's butchered infant. The problem of the bonobo mother Hide tearing her dead infant's forearm off and treating it like a chicken drumstick. The problem of the eight adult females, one juvenile female, and two infants who participated in the bonobo cannibalism at Kokolopori. The problem of very xenophobic naked mole-rats. The problem of Hölldobler's ant foreign policy of genocidal annihilation "whenever possible". Behaviours that are not a chance by-product of evolution by natural selection, but behaviours that are the very dynamics of evolution by natural selection. We now start to understand the paradox the rest of the book needs to deal with; of how to get beyond the problem of Humphrey, and the implications this will also hold for independent reason, non-contingent morality, human "nature" and extraterrestrial nature, and the evolution of machine intelligence. The problem of how to get beyond the blind, pitiless indifference of *both* transitions six and seven, and indeed every transition before these points. The problem of both getting to the eighth transition and what that eighth transition logically can, and cannot, then permit.

And we know there is, there has to be, an eighth transition for all three evolutionary traditions, albeit they offer different interpretations. For Darwin, we can never get beyond the problem of Humphrey, at least not at the biological level. Darwin's answer for the eighth transition was a second non-biological inheritance mechanism that can counter, albeit not always successfully, the first biological inheritance mechanism. The group-selectionists' answer for the eighth transition is the manifestation of an admittedly real, albeit very weak and easily subverted, form of natural selection in either one or, at most, a tiny handful of animal species. The sociobiologists' answer for the eighth transition is the appearance of a completely new way for natural selection to operate, and in just one unique species out of billions. Two of these traditions *do* offer hope to those "who do wish to find a basis for morality in nature and evolution", though Darwin himself does not offer a basis for morality – human or extraterrestrial – in biological nature. So far we have been allowing Darwin to speak first, and the group-selectionists and the sociobiologists to speak subsequently, but just for now we will reverse that inclination, because the latter two traditions offer us a simple, and breathtaking, conclusion to concentrate on, while Darwin's conclusion will be more nuanced and uncertain. *For the group-selectionists and the sociobiologists the vast majority of the intelligent extraterrestrials our species could, or will, ever encounter will be cannibalistic and homicidal.* Very xenophobic. Genocidally annihilistic. It may be somewhat ironic that by seeming to fight so hard to privilege the human animal genetically, the group-selectionists and sociobiologists appear to have doomed us to cannibalistic extraterrestrials, and xenophobic E.T.s, at least nine times out of ten, and indeed up to ninety-nine times out of a hundred. Let us develop this argument.

	Atypical group size	High intelligence
Darwin / Genic selection	Type II	Type II
Group-selectionism	Type II	Type I
Sociobiology	Type II	Type I

Remember that we are now solely interested in E.T., not multicellular life, not cosmic dinosaurs, and that we are defining E.T. as Brian Cox did: "when we talk about aliens ... we mean something that we can talk to", so something with a degree of intelligence. In the above table we are separating what from this point on we call "Type I" and "Type II", and the distinction is that Type I remains answerable to the processes possible under transitions six and seven, both non-colonial multicellular life and colonial, eusocial, living. Type I is both, is equally, Ntologi's chimp group and Hölldobler's ants. Whereas Type II is the eighth transition, be that the involvement of a second inheritance mechanism, natural selection starting to operate at the level of the group, or new mechanisms of natural selection never before encountered outside of the human animal.

For Darwin, who we return to below, real intelligence and very large and peacefully interacting groups come about together, as both are the result of transition eight. Both are Type II legacies. Yet the group-selectionists and the sociobiologists have either largely, or complexly, de-linked intelligence from the eighth transition. For both, the eighth transition, caused by an extraordinary genetic shift, is needed to explain very large and peaceful groups, but is not needed to explain high intelligence, which for them is a different genetic process. For both, peacefully interacting groups require Type II transition, but high intelligence is still largely, or at least partly, within Humphrey's Type I universe. Hence the group-selectionists argue that although individual-level and genic-level selection are predominant, in exceptional circumstances that have nothing to do with brain size genetic group selection comes to operate. Group-selectionists have in the past tried to attribute group selection to worker bees, lemmings, fish that are poisonous when consumed (where, it had been argued, "the toxin would be designed to destroy the enemies of the species", Williams 1966, p.228), harmful alleles in non-wild mice, naked mole-rats, chimpanzees (before the image of the playful, fruit-eating chimp "was gradually revealed to have a darker side") and bonobos, even if today many simply invoke it for human evolution. Which means that in, let's say, 99% of evolutionary cases, and de-linked from brain size, all the group-selectionists are left with is the "blind, pitiless indifference" of individual selection, be that Ntologi's chimp band or Hölldobler's ants.

Similarly, the sociobiologists aver that in one truly unique species out of billions, and seemingly largely de-linked from brain size, nature simply reversed the process of blind, pitiless indifference that it had been using for a billion years. The sociobiologists suggest mechanisms, or accidental evolutionary by-products, such as "metamorphosed" phenotypic expression, or Donald Symons' "*mal*adaptive" evolution, or "fictive ... faux-families", almost all of which seem to have nothing to do with a co-evolved large brain, and simply rely on a one-in-a-billion chance occurrence. Such as Helena Cronin's metamorphosed phenotypic expression where behaviour was "likely to change beyond recognition" once an animal leaves the grasslands. Even Robert Trivers' "strong selection" for a "multiparty altruistic system in which altruistic acts are dispensed freely", seemingly breaks, at least in part, the co-evolutionary link to a large brain. Because, as Richard Dawkins puts it, it is even possible that humankind's large brain "evolved as a mechanism of ever more devious cheating, and ever more penetrating detection of cheating in others" (1989, p. 188). So what both the group-selectionists and the sociobiologists are giving us is the potential evolution of high (or at least higher) intelligence in Type Is that in at least 90% – for the sociobiologists – and possibly even 99% – for the group-selectionists – of argued cases cannot be relied upon to have co-evolved with group selection or the reversal of the billion-year programme of pitiless indifference. For the two non-Darwin traditions, in 90% and possibly even as high as 99% of cases of extraterrestrial contact we will be meeting aliens that might be plenty smart enough to communicate with, to talk to, but they will also be murderous, xenophobic cannibals in Humphrey's mould.

Having considered the group-selectionist and sociobiologist answers, what of Darwin's answer? For Darwin, Type I cannot get us to either large and peacefully cohesive groups, *or* high intelligence, so Darwin offers us something very different from those at least nine murderous and cannibalistic intelligent alien species out of every ten. But what level of group co-ordination can Type I get to for Darwin, and what level of intelligence? Something perhaps more than just chimp social communication, basic emotional grunts and hoots, with more complex gestures and facial expressions? What about transition seven, then, and those colonial-living species? Yet our experience of colonial animals so far, from social insects to naked mole-rats, gives us nothing we can "talk" to, and nothing with any degree of real intelligence. Thus if Darwin's tradition is also to offer us Type I intelligence, we are probably going to have to hypothesise a form of extraterrestrial colonial creature that is smart enough for communication we can interpret and respond to, even if this falls a long way short of complex language. And we shall return to this interesting theoretical point when we consider machine intelligence, as there are already some thought-provoking "Type I" parallels, particularly in what is called evolutionary swarm robotics, where there appears to be evidence that communication and signalling can emerge spontaneously in the simplest robots, even when not explicitly programmed. So the question for Darwin's tradition is maybe not can such hypothetical biological creatures communicate, but can they communicate with us, and can they communicate with each other without possessing the second inheritance system interaction (Type II intellects) that Darwin moved on to, and that we discuss in the next chapter? Can they, in other words, remain Type I intellects, most likely colonial-living Type I intellects, and actually be something that we can talk to? The answer for Darwin's tradition seems to be scepticism; possible, but unlikely.

But even if we have found our first Darwin-tradition form of E.T., it is certainly a form that will never leave its home planet. They cannot visit us, or even have the technology to signal us; we will only know about this form of E.T. once we visit them. We started the book by touching upon what is called the Great Silence, and this is thus part of that silence. Even if a single inheritance mechanism can produce a level of Type I intelligence where we could have some form of basic communication, it appears, at least under Darwin's understanding, that it is never going to be enough to get them to industry or spaceflight, or even just radio. Which is obviously a good thing; if we do have a form of colonial-living E.T. that has achieved territorial dominance, this is still part of that universe of Type I blind, pitiless indifference; "ruthless and terrible predators". We will go to them, and they will look to kill us; they will have that "mystique of terror", and they will look to cut us to pieces where we meet. However, here we get to an irony of evolution, they will look to eliminate us, but they won't be up to the job. They will not have the technology, or the creativity. They will have the will to exterminate, but they will not have the tools to exterminate. If we go to them, they may murder us a few at a time, but they cannot kill us in our thousands.

If you want, or need, to kill tens of thousands of sentient beings at a time, if you want to be really *successful* at extermination, you need individuality, education, and a true division of labour where some are free to do nothing other than sit, think, and tinker, which is not achievable within Darwin's Type I universe. For Darwin, that will require Type IIs, and universities like Harvard, the self-same Harvard that developed, refined and perfected napalm across 1942, ready for it to be dropped on Tokyo on 10th March, 1945, burning alive tens of thousands of women and children. "I couldn't foresee that this stuff was going to be used against babies. ... That wasn't my business", were the later comments of Louis Fieser, the Harvard professor of organic chemistry who filed the patent (Neer

2011, p.283). This we can perhaps call the tragedy of Harvard; if you want to be truly good at extermination, at least within natural evolution and outside of machine evolution, you need Type II potential for both division of labour and advanced education. You will need more than just Type I "mobs of self-seeking individuals".

To kill us tens of thousands at a time you will need real intelligence, which at least for Darwin you can never get to in the natural world with a single inheritance mechanism. And it is worth confirming that it does not matter exactly how that single inheritance mechanism works. Maynard Smith and Szathmáry's third transition, which was only briefly mentioned in the first chapter, moved us from an undifferentiated RNA world where RNA acts as both gene and enzyme to the modern world where there is a division of labour, with the nucleic acids of RNA and DNA as carriers of information and proteins as enzymes. The RNA world hypothesis originated in the 1960s with scientists including Francis Crick, but the subsequent transition to the modern world would have allowed more efficient storage and transmission of information, including through the modern nucleic acid proof-reading process. (As also mentioned in the first chapter, some transitions are still under debate, and transition three is one of them, albeit all parties recognise the need for some intermediate step to the DNA world.) Maynard Smith and Szathmáry note, "in evolution, complex organs adapted for particular functions often arise in a simpler form, with a different function" (1999, p.42), where the process is driven by the same evolutionary imperative. But from a hypothesised RNA world to the modern division of labour, this is still Dawkins' process of "blind, pitiless indifference". Anywhere in the universe, and whatever the building blocks of later complex life, be that RNA, DNA, or some other inheritance system, with just a single mechanism we can only ever get to the same end-point, because we still have to deal with effective individual-level selection and all its implications for group size and group cohesion.

In the last analysis, or at least Darwin's final analysis, it seems the only real hope of meeting E.T. – certainly a spacefaring E.T., but even just a planet-bound E.T. that we can truly have a decent conversation with – is a dual inheritance mechanism. In the final evaluation, and under Darwin's own interpretation, the Great Silence is explained because we need yet another transition aside from the colonial living of number seven; we need that rare eighth transition, and we need Type II extraterrestrials.

THE THREE ROUTES ON TYPE I EXTRATERRESTRIAL INTELLIGENCE

For Darwin, then, and perhaps because of developments in evolutionary swarm robotics, the jury must still be out regarding whether Type I – transitions six and seven – could ever give rise to extraterrestrial intelligence we can communicate with, although it must be leaning heavily on acquittal or case not proven. For Darwin, we must wait for Type II and transition eight before we can reliably come across other-worldly natural intelligence that we can have a conversation with.

	Cosmic dinosaurs	Pitiless E.T. Type I
Darwin / Genic selection	✔	? ✗
Group-selectionism	✔	99% likely
Sociobiology	✔	90% likely

If Darwin is getting it wrong here, and the group-selectionists or sociobiologists are getting it right, the alien intelligence we will seemingly meet at least 90% of the time will be pitilessly indifferent to our future existence, and smarter than Darwin's Type Is can ever get to. For Darwin, high (book-smart) intelligence and large and peaceful group size come about together through the Type II overwriting of a susceptible brain. For the other two traditions large and peaceful group size is linked to a Type II genetic switch, but this is largely or wholly unconnected with higher intelligence, which remains possible even for their Type Is which have not gone through that eighth evolutionary transition. Hence their Type Is can be smarter than Darwin's Type Is, but not napalm-smart, because napalm-smart still requires Harvard and very large group sizes, which is their Type II evolutionary transition. Their Type Is can be communication-smart, but not book-smart or engineering-smart, because they will have no books and no engineering, as books and engineering require advanced co-operation and a complex division of labour. Their Type Is will still need to kill (and eat) us dozens at a time, not thousands at a time. Fanatically homicidal, but unlikely to be fanatically genocidal.

Yet for those who want or need their avuncular E.T.s warping around the universe bestowing unconditional largesse, do understand that only the group-selectionists and the sociobiologists can ever give us this, because while rootlessness and a degree of irrationality will be Darwin's Type II, avuncular E.T. becomes the group-selectionist and sociobiologist second form of biological intelligent life, their Type IIs. And where that genetic switchover has given us a Kropotkin-like "basis for morality in nature and evolution", albeit a very rare Type II, as for the group-selectionists homicidal Type Is make up 99% of intelligent alien species, as intelligence has been wholly de-linked from group selection. Or to put it another way, the group-selectionists need two serendipitous biological transitions to get to a wholly benign E.T., being both large brains *and* the switch to group selection, and such serendipity will only happen at most one time in a hundred. For the sociobiologists, homicidal Type Is make up at least 90% of intelligent alien species, as intelligence has been largely, but not wholly, de-linked from benign biology. Helena Cronin's metamorphosed phenotypic expression, for example, wholly de-links from brain size, as does

Donald Symons' maladaptive evolution. Yet Steven Pinker's "cognitive twist" and "environmental cues", where we come to falsely view complete strangers as kin, might just be argued to be a function of a larger brain, even if a seemingly idiotic and easily fooled larger brain, albeit it is worth flagging ahead of the coming chapters that both the group-selectionist and the sociobiological traditions seem to build their Type II genetic architecture from wholly unrealistic and ahistorical assessments of human rationality and human morality.

Darwin could explain the Great Silence because high intelligence, including radio-level and engineering-level intelligence, needed the rare eighth transition through Type II overwriting of a susceptible brain. Under the group selection and sociobiology models we can possibly still explain the Great Silence, because for them homicidal Type I extraterrestrials predominate, and though capable of being smarter than Darwin's Type Is they will still lack the level of intelligence needed to leave their home planet, or even communicate electronically, while for the group-selectionists and sociobiologists their avuncular space-faring E.T.s are much, much rarer, requiring both the evolution of higher intelligence and the quite separate move through that uncommon transition eight. Still, it is an extraordinary realisation that these two alternate traditions of modern evolutionary biology offer us only the remotest possibility of extraterrestrial intelligence ever being more than homicidally, pitilessly indifferent to our very existence.

It may seem at first sight surprising that, given Darwin and the alternative two traditions have such very different models of natural intelligence, they can still give us identical conclusions in the other key areas. So all three give us cosmic dinosaurs, and all three will give us truly genocidal machine intelligence. At first sight surprising, but perhaps less so once we recognise that all three traditions are ultimately traversing the scholarly minefields of human intelligence and human social and economic organisation. As Darwin wrote to Wallace in December 1857, and well ahead of publishing his 1871 *Descent of Man*: "You ask whether I shall discuss 'man'; I think I shall avoid the whole subject, as so surrounded with prejudices" (Raby 2001, p.134). It is only on the question of natural intelligence that Darwin and the other two traditions diverge wildly. Darwin's answer is, however, more nuanced, more messy, less black and white, more shades of grey. Darwin can't give us either the extraterrestrials of our dreams, or the extraterrestrials of our nightmares.

For Darwin, Steven Spielberg's E.T. won't exist. For Darwin, we can get to a race of intergalactic botanists, sure, but they will be following the Holy Scriptures, which tells them of their sacred duty to sanctify plant life, even as they remain largely unmoved about violently ending highly intelligent non-plant life. Or for Darwin we can have one small spacecraft of gentle interstellar gardeners, but they will be seeking time away from their planet chock-full of conspiracist and intolerant leathery compatriots. For the other two traditions, though, there can be only the black and the white. From them we get the nightmares plus the dreams. Isn't it mind-boggling to realise that if Darwin was in fact wrong, and the other two schools are instead correct, then nine times out of ten where we will encounter higher intelligence, even ninety-nine times out of a hundred, we will find a planet teeming with canny, homicidal extraterrestrials? The savage hunters of James Cameron's *Alien* perhaps, that wonderful transition seven re-imagining, implanting embryos inside their living human hosts in a scaled-up version of the parasitic wasp, one of the terrestrial examples of koinobiont parasitoids. Humphrey, Hide, and Ntologi, but smarter, more dominant, and even more ruthless. If Darwin was wrong on this point, then over the longer term astronauts are going to want one heck of a pay rise. But one time out of ten, one time out of a hundred, the group-selectionists and the sociobiologists can still give us the polar opposite that many will always yearn for; the race of gentle, kindly, and ecologically wise galactic botanists spreading cosmic harmony and the secrets of interstellar engineering.

4

E.T. TYPE II – A DUAL INHERITANCE MECHANISM

> "Let us understand, once for all, that the ethical progress of society depends, not on imitating the cosmic process, still less in running away from it, but in combating it. ... The history of civilization details the steps by which men have succeeded in building up an artificial world within the cosmos."
>
> – **Darwin's bulldog** (Huxley 1894, p.141)

Before we turn in detail to defending Darwin's Type II extraterrestrial intelligence, let us recap. We have seen that all of the traditions of modern evolutionary biology give us a galaxy almost certainly teeming with dinosaurs, because all of the traditions put something very unusual and very rare happening after, and not before, transitions six and seven: the move to complex multicellular life and the move to eusocial colonial living. Each of the three traditions also offer us homicidal, cannibalistic Type Is as the only possible form of intelligent extraterrestrial life *unless and until* we can trigger that very unusual and very rare further move to transition eight. Each of the three traditions needs an eighth transition to get to very high intelligence, even if the three traditions offer us somewhat different interpretations on that eighth transition.

	Requires a change to the template of evolution?	Fails the parsimony test?	Invokes new evolutionary processes?
Darwin / Genic selection	NO	NO	NO
Kropotkin / Group selection	YES	YES	NO
Kelvin / Sociobiology	YES	YES	YES

For Darwin, peacefully interacting large groups and real deliberation requires a second inheritance mechanism that can exist at odds with the rules generated by the primary inheritance mechanism. Exist contrary to what Darwin's bulldog T.H. Huxley called the "cosmic process" of evolution by natural selection that all other life anywhere must answer to, and what the philosopher of science Dan Dennett ninety years later would term Darwin's "universal acid". For Darwin, there wasn't a, there could be no, biological answer to the problem of Humphrey, though the other two traditions disagree. The other two traditions offer us a vast genetic switch to Type II existence, not Darwin's language-based, non-genetic one. So the group-selectionists are citing the move from individual selection to group selection sometime after we broke from a common ancestor with the chimpanzee. That would have to be a very rare biological transition; a change to the major behaviours template that evolution had seemingly exclusively worked with for a billion years. The sociobiologists are invoking new evolutionary processes, such as third-party reciprocal altruism and "fictive ... faux-families", never before seen or needed. That would have to be a uniquely rare biological transition; not just a change to the predominant behaviours template, but a change to the very processes evolution had exclusively worked with for a billion years. So all of the traditions of modern evolutionary biology need an eighth transition, and the move to Type IIs, to explain real empathy-based intelligence, human or extraterrestrial, and a transition that effectively applies to only one species out of the few million animal species alive today on this planet, and to only one species out of the billions of animal species that have ever lived here.

But for reasons we touched upon in the first chapter, we have to conclude for just one of the three traditions, so please understand further that both group-selectionism and sociobiology are again failing to act parsimoniously in the explanation of empathy-based deliberative life. Because both are invoking two changes at the eighth transition, whereas Darwin is invoking just one. Or to put it another way, both are effectively positing not just an eighth transition, but a simultaneous and unrelated ninth transition. Darwin saw the eighth transition to be the evolution of the capacity for language, acting on a very large and psychologically-susceptible brain (and both are co-evolutionary; language requires a large and densely synapsed brain, and a large and densely synapsed brain permits language – across twenty-five sampled species of primates "humans displayed the highest number of synapses per neuron", Sherwood *et al.* 2020, p.5604). Sociobiologists see the eighth and ninth transitions as being the rewriting of nature's billion-year biological code of blind, pitiless indifference *and* the simultaneous and seemingly unrelated evolution of a large and densely synapsed brain and the unique capacity for language. Group-selectionists see the eighth and ninth transitions as being a move to group selection *and* the simultaneous and completely unrelated evolution of a large and densely synapsed brain and the unique capacity for language. Highlighting two transitions doesn't of course make those proposals wrong, but it does again make the proposals less likely, less parsimonious.

For both of the non-Darwin / non-genic selection traditions a vastly rare and unique transition, a seemingly one-in-a-billion-species transition, was coincident with two significant changes, while Darwin required just one of those two changes to explain the same event, and indeed required only an evolutionarily much more plausible and simple development. But it is not just a failure to act parsimoniously, because we should now highlight that even the current crop of sociobiologists accept not only that culture is enormously powerful, but that culture must be able to actively override our deepest genetic impulses.

	Common ancestor	Early humans	1. Good & 2. Evil	Changes required
Darwin	Genetically immoral	Genetically immoral	1. Culture 2. Genes/Culture	Good: 1 Evil: 0/1
Group-selectionism	Genetically immoral	Genetically moral	1. Genes 2. Culture	Good: 1 Evil: 2
Sociobiology	Genetically immoral	Genetically moral	1. Genes 2. Culture	Good: 1 Evil: 2

The 1992 book *The Adapted Mind* is seen as the seminal work of contemporary sociobiology. And to quote Cosmides, Tooby and Barkow's introductory chapter: "The central premise of *The Adapted Mind* is that there is a universal human nature, but that this universality exists primarily at the level of evolved psychological mechanisms, not of expressed cultural behaviors. On this view, cultural variability is not a challenge to claims of universality, but rather data that can give one insight into the structure of the psychological mechanisms that helped generate it" (p.5). In other words, sociobiology, today sometimes called evolutionary psychology, holds that our similarities are genetic, but our differences are cultural, and *Scientific American*'s John Horgan cites their literature. "'Evolutionary psychology is, in general, about universal features of the mind,' they have written. 'Insofar as individual differences exist, the default assumption is that they are expressions of the same universal human nature as it encounters different environments'" (1995, p.153). Ok, but what does this actually mean?

Central to the programme of human sociobiology is the conception that humanity is, biologically, "*the* moral animal" (Wright, 1994), that "morality evolved as instinct" (Wilson 1978, p.5), that human phenotypic expression left us with a "font of *vast* altruism" (Cronin). Sociobiology has human DNA transitioning from immorality (or at least Humphrey's pitilessly indifferent amorality) to morality at most a couple of hundred thousand years ago. But look at the above table: this leaves sociobiology with a real problem, because it now has to explain how "the" moral animal has spent much, if not most, of its recent history as a brutal thug. So how does sociobiology explain human large-scale evil; the pogroms, the mass rapes, the slavery, the segregation, the genocides? It invokes culture. We are, sociobiology says, genetically and uniquely, programmed to be largely moral and good, but culture can distort this and take us back to immorality and evil. So this leaves sociobiology with two additional problems. The first is it is once again breaking the principle of parsimony. Sociobiology requires two steps to explain evil – a first transition to moral DNA, then a second step where culture overwrites that benign genetic coding to get us back to nation-level evil. Yet Darwin required either zero steps – evil is nothing more than a species-wide mammalian evolutionary legacy – or at a maximum just one – we may be born pitilessly indifferent, but human large-scale immorality is not indifference and involves creatively planned cultural nastiness. In contrast with both of Darwin's possible offerings sociobiology is adding additional steps. Sociobiology's second problem is it is already accepting the enormous power of culture to overwrite biology, and if you are going to do this there was no point positing in the first place a seemingly impossible vast genetic transition, or seemingly improbable vast "*mal*adaptive" evolutionary transition, and once we left the savannas and the grasslands. If you are anyway going to accept the power of culture to overwrite biology it is far more reasonable to do it without first positing a breaking of the billion-year evolutionary mould. Human sociobiology is not just historically and mathematically wrong; it seems to be self-underminingly wrong.

But the group-selectionists fare no better, which shouldn't really surprise as Maynard Smith always said that human sociobiology was just another form of confused group-selectionism, and Dawkins that "frequently the evolutionary preconception in terms of which such [sociobiological] theories are framed is implicitly group-selectionist" (1989, p.191). Group-selectionists, generally more on the political left than the sociobiologists, openly admit the power of culture to distort genetic imperatives, but there is again no need to invoke group selection if you are already going to make such an admission. The group-selectionists argue that there needed to be no sociobiological genetic or phenotypic transition from immorality to morality a couple of hundred thousand years ago on the grasslands, but only because through group selection we evolved benevolence, Kropotkin's "feeling infinitely wider than love or personal sympathy", at an even earlier stage. But this just pushes the problem back further in time; it doesn't resolve the problem.

Group-selectionists do not attempt to attribute morality to trilobites or crocodile-line archosaurs, and most group-selectionists accept with de Waal that our closest cousins the chimpanzees live in "a world without compassion". This means that even within this tradition group selection with large-scale effects probably wasn't operating until after we split from a common ancestor with chimps and bonobos. A couple of decades ago group-selectionists might have tried to argue that humans were not actually unique, and that bonobos had evolved benevolence too, and thus that group selection may have operated ahead of the split from a common ancestor, but then had been subsequently lost in chimps. However, as previously discussed, including recalling Hide's novel use of an infant's hand and forearm, the last decade or so has generally put paid to the idea that bonobos escaped from that world without compassion. Be that as it may, at some point ahead of the evolutionary switch posited by the sociobiologists the group-selectionists have to theorise a historically amoral, compassion-less, process switching across to benevolence and a feeling infinitely wider than love. And then *subsequently* they are invoking the power of culture to explain human large-scale evil. Again, we have additional unnecessary steps, alongside an open admission of the power of culture. This also makes group-selectionism not just historically and mathematically wrong, but self-underminingly wrong. Now, we dealt with the group-selectionist and sociobiologist conclusions for extraterrestrial Type IIs in the last chapter, those wise and kind-hearted, though very rare, E.T.s warping around the universe bestowing ecological sanity and unconditional assistance. But we have yet to deal with Darwin's own Type II extraterrestrial, which is what we turn to now.

> "Civilized human behavior has about as much connection with natural selection as does the behavior of a circus bear on a unicycle."
> – **Mark Ridley** and **Richard Dawkins**, "The natural selection of altruism"
> (1981, p.32)

In his classic 1966 work George Williams provides us with a wonderful example of the utter incongruity between human sentimentality and Dawkins' natural-world "pitiless indifference".

E.T. TYPE II – A DUAL INHERITANCE MECHANISM

He recounts the attitudes of an audience being shown a film about elephant seals: "Amid the crowded but thriving family groups there was an occasional isolated pup, whose mother had deserted or been killed. These motherless young were manifestly starving and in acute distress. The human audience reacted with horror to the way these unfortunates were rejected by the hundreds of possible foster mothers all around them" (pp.188–9). Williams' point had been to show the fallacy of good-of-the-species theorising – "It should have been abundantly clear to everyone present that the seals were designed to reproduce themselves, not their species" (p.189) – but the passage serves equally well to demonstrate the leap of faith it takes to suggest that the blind, pitiless indifference of natural selection could have evolutionarily morphed into common human decency.

When the sociobiologist Sarah Hrdy was documenting the casual slaughter she witnessed in langur monkeys, she admits that it was her own tears that were among the problems she had to overcome. And Takayoshi Kano, leader of the bonobo research project at Wamba, recorded that when his colleague Mariko Hiraiwa-Hasegawa observed male chimps in the process of tearing an infant from its mother "Hasegawa momentarily forgot her position as a researcher and, brandishing a piece of wood, she intervened and confronted the males to rescue the mother and infant" (cited in De Waal 1997, p.119). Note the danger that Hasegawa was knowingly putting herself in here: one human female confronting multiple violent adult male chimpanzees each of which, Hasegawa would have been fully aware, more explosively powerful and faster than the average human adult male, because while we have more slow-twitch muscle fibres for endurance they have fast-twitch fibres for speed and power. While *every* adult langur or chimpanzee will tear infants of their own species to pieces without batting an eyelid, humans shed tears over the deaths of infants of *another* species, or run in to protect them against attacks by their conspecifics even at significant risk to life and limb, or display both written and verbal warnings before showing footage of a mother bonobo using her dead infant's forearm as a cocktail stick. And this is supposed to be a fundamentally new evolutionary programme that has evolved once in all the billions of years of evolution on this planet, and once in billions of species? For reasons we will come on to, it cannot be the case that individual-level or gene-level natural selection has come up with a unique and antithetical genetic strategy for humankind. DNA neither knows nor cares. DNA just is.

A single inheritance mechanism cannot give us either civilised behaviour or the group sizes necessary for education, advanced learning and deliberative intelligence. And thus the sociobiology notion that we "out-evolved" a billion-year pattern of pitiless indifference is plain wrong. Wrong mathematically, as we saw in the second chapter, but also wrong practically. It cannot explain human behaviour because, as Maynard Smith notes, "human societies change far too rapidly for the differences between them to be accounted for by genetic differences between their members" (1992, p.82). Or as Richard Dawkins writes: "Many human societies are indeed monogamous. In our own society, parental investment by both parents is large and not obviously unbalanced. ... On the other hand, some human societies are promiscuous, and many are harem-based. What this astonishing variety suggests is that man's way of life is largely determined by culture rather than by genes" (1989, p.164). The arguments of those trying to justify human behaviour in genetic and adaptationist terms "do not begin to square up to the formidable challenge of explaining culture, cultural evolution, and the immense differences between human cultures around the world" (p.191). But wrong also because nature cannot and does not work this way. Nature cannot make a leap. *Natura non facit saltum.*

"On the theory of natural selection we can clearly understand the full meaning of that old canon in natural history, 'Natura non facit saltum'," Darwin wrote (1859, p.233). The Latin expression quoted by Darwin in the sixth chapter of *Origin* was from Linnaeus' classic 1751 work on taxonomy. Darwinian evolution can accommodate ideas like genetic drift, developmental plasticity, and epigenetic – literally "on top of" the genetic – inheritance, but saltation theories are theories that rely on macromutation, or a sudden large beneficial jump that is consequently incorporated into the gene pool of a species. Saltationism is a neo-Darwinian heresy. And yet sociobiology's claims are not simply implicitly group-selectionist involving processes impossible under individual-level Darwinian inheritance, they are also saltationist. The suggestions would fundamentally have to rewrite natural selection's genetic code, being the immorality detailed above, the world without compassion of chimpanzees, bonobos and our other close kin. Sociobiology accepts that up until 100,000 or so years ago our ancestors were selfish apes. If sociobiology were to be right, what would have to be the implications for that last genetic change?

All primates, all mammals, all vertebrates, all animals, are born programmed through an identical behavioural code, the code Richard Dawkins terms pitiless indifference. This codes for species-wide patterns of cannibalism, infanticide (and the voluntary mating with those that have killed your infant), rape, levels of lethal violence against same-species members many thousands of times higher than rates found in even the worst cities, and indifference to the suffering of non-kin. Patterns not that *some* within a species are coded for, but that *all* within a species are coded for. This code is not restricted to a few murderous genes; gross immorality is the programme of natural selection. Utter selfishness and indifference to others' suffering (where they can be of no use to you, or are not closely related to you) is a message coded into the entire natural world behavioural genome. It has been the guiding force behind evolution for over a billion years, because it works very efficiently. Nature has taken millions of years to code for the set of behaviours in primates, and while you can tinker with these behaviours, and add to them, and remove certain behaviours gradually over time, you cannot rip up the rulebook and start again.

The anthropologist Jared Diamond famously described humans as the third chimpanzee in his book *The Rise and Fall of the Third Chimpanzee*. Humans share around 99% of their DNA with both the common chimpanzee (*Pan troglodytes*) and the bonobo (*Pan paniscus*), once called the pygmy chimp. We split off from a common ancestor with the chimpanzee and the bonobo some six to eight million years ago. The point of the first part of Diamond's book was to show – using the work of the molecular biologists Charles Sibley and Jon Ahlquist (1984) whose studies have since been reconfirmed by many others – that human beings are actually more closely related genetically to both species of *Pan* than both species of *Pan* are to the other apes. As Diamond noted, the common ancestor to the chimp, the bonobo and the human split off from the ancestor to today's gorilla, the next nearest ape, more than a million years before the common ancestor to chimps, bonobos and humans split. In consequence, we and the chimp and the bonobo all share just less than 98% of our DNA with the gorilla. And the genetic distance separating humans from bonobos or chimps (less than 1.5%) is actually less than the genetic distance between the common gibbon and the siamang gibbon, who were found by Sibley and Ahlquist to have a variance above 2%. The human/chimp relationship is usually given as just under 99% when calculating using precise alignment and thus single nucleotide substitutions, and although this method has been standard when comparing human/chimp DNA, it can be slightly misleading. For example, in September 2002 the molecular biologist Roy Britten reported that the human/chimp relationship falls to perhaps 95% when one also includes indels, insertions and deletions of DNA found in one species but not the other. There is some dispute about the significance of such findings, although others have confirmed this indel nuance (see Varki & Nelson, 2007, and calculating a 96% overall correlation), but since indels exist across other species boundaries it can never really change the observation that humans are more closely related to both species of *Pan* than both species of *Pan* are to the other apes. Our DNA is remarkably close to all other apes, and the idea that this 1.5% difference – this difference gained in the small evolutionary period since we split from our common ancestor – can fundamentally rewrite the rules of evolved behaviour seems just wrong.

But the problem is worse than that. As Diamond went on to explain, although we split from a common ancestor with the chimpanzees some six plus million years ago "for most of the time since then, we have remained little more than glorified chimpanzees in the ways we have made our living" (1991, p.27). Our ancestors remained little more than glorified chimpanzees, with all that that should imply, until a couple of hundred thousand years ago. And these glorified chimpanzees were genetically even closer to us until, in the last instance, "perhaps they shared 99.9% of their genes with us. ... The missing ingredient may have been a change in only 0.1% of our genes" (p.46). There is absolutely no evidence that these glorified chimpanzees, several distinct species of *Australopithecus*, *Homo habilis*, *Homo erectus* and even archaic *Homo sapiens*, managed to break from the pattern of pitiless indifference – the "world without compassion" of the chimpanzee, the "eight adult females, one juvenile female, and two infants participated in the cannibalism" of the bonobo – that runs throughout single inheritance system natural selection. These ancestors, finally only maybe 0.1% away from modern humans in their genetic makeup (such a tiny variance that this same 0.1% is the genetic distance between two modern human beings selected at random) were still coded for cannibalism and infanticide. *And sociobiology fully accepts this,* which is why sociobiologists postulate that morality emerged from immorality some 100,000 or so years ago in those last few genetic blips – or favourable maladaptations and serendipitous phenotypic expressions – which produced modern *Homo sapiens*. Yet, even ignoring the non-existence of such a selecting mechanism, it seems an evolutionary impossibility that we could have retained 99.9% of a chimp's DNA, glorified or otherwise, yet expect that final infinitesimal 0.1% to have been the locus of the drive for indifferent selfishness as old as life itself. Instead, for Diamond, as for Darwin and the gene-selectionists, something *extra* was added in that last 0.1% change. We didn't *lose* anything, and certainly not an almost four-billion-year pattern for extreme selfishness, we *gained* something. We gained a susceptibility to culture with the emergence of the capacity for spoken complex language, although this susceptibility required a very large and densely synapsed brain which had itself taken many hundreds of thousands, even millions, of years to evolve. Evolutionary geneticists have long known that these last tiny changes could have easily added the necessary genes, and that both speech and brain wiring complexity may depend on very few new genes. The reasonable view is that we received something extra in that last 0.1% shift. Just compare this view with the sociobiology (and, indeed, group-selectionist) alternative that nature backtracked through the human genome, deleting or inhibiting genes for the great majority of animal behaviours while substituting genes for a vast number of new and opposing human behaviours, or their other view that we retained 99.9% of a cannibal's DNA yet "our modern environment" metamorphoses "beyond recognition" (Cronin) a billion-year behavioural archetype.

Mutations with large effects, or macromutations, do occur, and where they produce a radical difference in the visible characteristics shown by an organism such is termed a "monster". But macromutations cannot contribute to evolution because they will be eliminated by natural selection. The great majority of mutations are deleterious to offspring, and are consequently removed by selection. And as Dawkins puts it, "The greater the number of simultaneous improvements we consider, the more improbable is their simultaneous occurrence" (1986a, p.234). Most complex characteristics under genetic control are, and have to be, polygenic, governed by the combined interaction of a number of genes. So if you argue that in that last genetic spurt humankind inherited genetic coding for all sorts of new moral and behavioural attributes you are positing the chance addition of multiple new (working) gene complexes. Whereas changing brain size several times since departing from that common ancestor with the chimp, or modifying synapse densities and dendritic branching; here you are only adding in stages to what already exists, just tinkering with what is already there in a very straightforward, albeit somewhat time-consuming way. In contrast, sociobiology's theories would require entirely new fully wired-up behavioural mechanisms, even the foundations of which could not conceivably have existed in our ancestors, and the removal of existing and largely antithetical behavioural coding. We are thus into the realm of astronomical statistical improbability. "Darwin was a passionate anti-saltationist, and this led him to stress, over and over again, the extreme gradualness of the evolutionary changes that he was proposing", notes Dawkins (p.248).

A LOST AND ROOTLESS OUTSIDER

For Darwin, evolutionary logic *forces us* to acknowledge a second inheritance mechanism that must at least in certain key respects be able to counter our first inheritance mechanism, although it would be another century before the giants of modern evolutionary biology – names like George Williams, John Maynard Smith, Bill

Hamilton and George Price – would provide the mathematics to prove Darwin and Huxley correct. "A belief constantly inculcated during the early years of life, whilst the brain is impressible, appears to acquire almost the nature of an instinct", wrote Darwin (1871, Pt. i, p.100). Or as Maynard Smith put it, "once a species has acquired language as a second method of passing information between generations, new mechanisms of change become possible" (1993, p.43). Although nature had given us certain traits which could be built upon, or twisted, something else had to be doing the building and the twisting. How could truly virtuous human self-sacrifice, where, said Darwin, the bravest men "freely risked their lives for others" (1871, Pt. i, p.163), come about? Ultimately, Darwin said, the social virtues could only be explained as cultural, not biological. Guided by the approbation of our fellow men, ruled by deep religious feelings, confirmed by instruction and habit, "all combined, constitute our moral sense or conscience" (Pt. i, p.166).

As Maynard Smith says of altruism, biologists "have explanations – such as the fact that the altruist may share genes with the recipient of its altruism, and it is genes, not individuals, that matter in evolution – but they are ones that work only for altruistic behaviour among the members of small groups" (1992, p.119). As with morality, so for Darwin with human group sizes. De Waal writes: "What is most amazing is that our species is able to survive in cities at all, and how relatively *rare* violence is" (1996, p.195). Chimps can live only in groups of up to one hundred or so individuals, and bonobos tend to live in similar or slightly smaller groups. Yet humans live side by side with non-kin in cities and nations of millions, and subscribe fanatically to vast religions, group sizes unknown outside the immense kin groups of the social insects. The philosopher of science Elliott Sober and the biologist David Sloan Wilson are no friends of selfish gene theory (e.g. Wilson, 2015). Yet they openly admit the logical and mathematical incongruity of humans being able to live in cities. Human groups have, like social insect colonies, "been interpreted as superorganisms for centuries", say Sober and Wilson in their book *Unto Others*, but biologists need some explanation for "why humans are ultrasocial" (1998, p.158). D.S. Wilson is still one of the most influential levels-of-selection theorists within what is known as the multilevel selection (MLS) tradition. Sober and Wilson find their answers to human cohesion in genetic group selection; as they put it: "At the behavioral level, it is likely that much of what people have evolved to do is *for the benefit of the group*" (p.194; emphasis theirs). Similarly Martin Nowak – who wrote the 2010 paper with Ed Wilson attacking Hamiltonian kin selection and placing eusociality under multilevel selection and mutations prescribing the persistence of the group – calls humans "SuperCooperators" (2011, writing with the *Telegraph* science editor Roger Highfield). In contrast, for Darwin, human group size cannot be explained by D.S. Wilson's group-selectionist theories, and cannot be explained by sociobiological theories that humans live in vast "fictive ... faux-families" (Pinker, 2012), or live with "metamorphosed" phenotypic expression (Cronin 1991, p.329). For Darwin, it is only language operating on a large brain that can redirect, that can reprogramme, the cosmic process.

We now understand why Darwin concluded as he did, with all the implications this automatically has to hold for extraterrestrial intelligence, but we are still left with a significant epistemological problem that refuses to go away. A single inheritance mechanism is governed by the blind, pitiless indifference of the DNA world, that "world without compassion". A single inheritance mechanism cannot get us to compassion or, at least in diploid vertebrates, to the large group sizes and division of labour necessary for book-smart advanced thinking and technology. A second inheritance mechanism including culturally-enhanced napalm-smart memory can give us the necessary intellectual flexibility, by opening us to that inner world, that self-monologue, that theory of mind (ours and others'), opening us to ideologies and concepts, to the echo-chamber of both fears and fantasies. Driven by, and driving, language. The lingering epistemological problem, though, is that we have largely escaped the utter ruthlessness of one inheritance mechanism, but seemingly only by leaving ourselves at the mercy of contingent external ideas and knowledge. The contingency of that second inheritance mechanism, but also the contingency that comes with having two largely incompatible programming systems. Two different and even warring operating systems, if you like. Because there is now no way to get "outside" the incompatible programming. Sure, a second inheritance mechanism may have allowed vast cohesion and all of human civilisation, but that doesn't automatically make everything that comes with it just or right, or mean we can even recognise truth or fairness. Group-selectionists and sociobiologists may think they can give us avuncular Type II extraterrestrials warping around the universe imparting unreserved generosity (albeit alongside their far greater number of Type I cannibalistic extraterrestrials), but Darwin's Type II is much more nuanced and fragile than this.

Everything else in nature is an *insider*. Programmed solely through the pitiless logic of DNA, or at least a single inheritance mechanism. Party to a ruthless one-stop agenda. For Darwin, we are, and any Type II extraterrestrial will be, an *outsider*, and a largely lost and rootless outsider. Sitting partly outside the billion-year genetic agenda, answerable now to two largely incompatible agendas. Unless reason, human reason or extraterrestrial reason can somehow come to our, or their, aid, can somehow get us, or them, to the mental quietude wherein lies objective right and truth – and we will be moving on to examine this possibility in the next chapter – we Type IIs are storm-tossed in the way no other creature can ever be. Type I involves, as Dawkins put it, "no evil and no good, nothing but blind, pitiless indifference". We have thus gone beyond the relentless amorality of the single inheritance mechanism. Both good and evil may now be open to us, but unless reason is an end in itself, and unless we can achieve that critical level of rationality, will we ever have the capacity and the self-knowledge to truly know one from the other?

"Most people are other people. Their thoughts are some one else's opinions, their lives a mimicry, their passions a quotation", wrote Oscar Wilde (1905, p. 59) in 1897. Yet Wilde is in one sense quite wrong. We are all other people, or at least complex amalgams of other people, not just most of us. We are all created by someone else's culture working on top of a species-wide, billion-year, biological pattern. In a universe where there can be no freedom of choice – that "general delusion", as Darwin called it (Barrett *et al.* 1987, p.608), that "stupid idea", as George Williams termed it (letter, 27 November 2002) – and where biological coding was effectively put in place hundreds of millions of years ago, yet where reason cannot, or does not, ride to our rescue, then we are all "other people". This is not necessarily something to fear, but it is something to understand, something that we need to learn to accept. As Nietzsche once put it, to want to be the cause of oneself, what is termed the *causa sui*, "is the best self-contradiction hitherto imagined, a kind of logical rape and unnaturalness … nothing less than the desire… to pull oneself into existence out of the swamp of nothingness by one's own hair" (1886, pp.50–1). Though we can still take comfort that it is false to think that determinism – the understanding that all current and future events are necessitated by past ones – and novelty are incompatible. Given the massively complex interaction of deterministic forces that build upon each other, never-before-seen configurations are inevitable and everywhere, and we are part of an unfolding process of extraordinary originality. We can also take comfort that it is false to think that determinism implies a fatalist resignation to events. The British philosopher Derek Parfit showed that Kant was wrong to imply that determinism would force us to be *passive*.

"Even if determinism is true, we can be *active*, by trying to make and to act upon good decisions", writes Parfit (2011, p.262). Actions and efforts still have effects and change the outcome from what it would have been if the effort had not been made; it is just that the outcome can still be defined in terms of prior causes. Sometimes, and notwithstanding a life deterministic at the human level, the world can move on when someone makes the effort to write a book explaining what Darwinism actually implies about both natural intelligence and non-natural, machine, intelligence.

We are Darwin's second form of E.T., we are his Type II, we are his truly intelligent extraterrestrial, and Type II extraterrestrials will always be like looking in a mirror. We shall know when and if we can trust them by examining when and if we can trust ourselves. Trust ourselves to tell the truth, but even before that trust ourselves to understand the truth. We shall know when and if they can be moral by examining when and if we can be moral. It is quite extraordinary to realise that although we cannot know for certain if that neighbouring fence will survive the next strong wind, thanks to Darwin we can know with absolute certainty the history and psychology of true intelligence anywhere in the universe, even though it must exist trillions upon trillions of miles away. A history and a psychology set in motion across billions of years by the relentless, and constrained, logic of Darwinian individual-level selection. They will have their gods, their prophets, their holy wars, and their conspiracy theories. They will have seen pogroms, caste systems, slavery, segregation, ethnic cleansing and genocide, even if the historical details may be somewhat different from ours. Because on their worlds maybe blue belittles green, and sexual dimorphism has left females as the larger and more aggressive gender. If so, perhaps on these worlds adult females genitally mutilate young males without any medical justification, while half their planet refuses to outlaw the practice. On these worlds, maybe adult males are viewed as second-class citizens, or not treated as persons in their own right, or even as the property of their wives, mothers, sisters. And we know they will have tried theocracy, autocracy, monarchy, oligarchy, and aristocracy, and if we are very lucky they may have occasionally got to democracy, although many of their vastly influential super-wealthy will nevertheless be actively working to undermine civil society and representative government. But they will also have compassion, love, kindness, courage, honour, and rudimentary ethical and educational systems.

Because we cannot begin to understand human psychology, or human history, without recognising such contingent mimicry, such complex amalgams of other people. Abraham Lincoln may have freed the slaves, may have been among the finest Americans of his generation, yet he was still a nineteenth-century bigot, and white supremacist, at heart. In September 1858, just a few years ahead of his Emancipation Proclamation, he said the following: "Ladies and gentlemen ... I am not, nor ever have been in favor of bringing about in any way the social and political equality of the white and black races. ... There is a physical difference between the white and black races which I believe will for ever forbid the two races living together on terms of social and political equality. ... While they do remain together there must be the position of superior and inferior" (Lincoln, 1858). And while his views did ameliorate to a small degree before he died, and over the single political issue of – very limited – black male suffrage, he never withdrew his comments that black people should not be allowed by law to intermarry with whites, or serve on juries, or hold office. Lincoln, Honest Abe or the Great Emancipator, a man known for his humility yet incongruously memorialised in Washington as 28 feet of white marble, may have been a fine man by the standards of his day, but he was still the product of his time and his place, and incapable of escaping the intellectual box of his upbringing. Because Lincoln exemplifies what will always be the problem with Type IIs, the storm-tossed outsiders. Finally capable of good and evil, but now in the same person, and at the same time. Liberator and bigot. Redeemer and fool. Mimicry without self-evaluation, and largely without self-awareness.[3]

There is something called the Kardashev scale of galactic civilisation, often cited by science writers like the American theoretical physicist and futurist Michio Kaku. We are apparently on our way to one day becoming a level-one Kardashev scale civilisation, to being capable of making use of all of the energy that radiates to us from the Sun rather than, as at present, just a tiny fraction of it. Level-two Kardashev civilisations are defined as capable of making use of all of the energy radiating from their star, not just that proportion falling on their home planet, and level-three civilisations are defined as way beyond even this. In fact, some already argue that we have glimpsed the enormous power of Kardashev-scale civilisations. Abraham (but known as Avi) Loeb is the Harvard astrophysicist and black hole specialist who has argued that the somewhat anomalously-shaped and moving, and unusually reflective, 'Oumuamua space rock which recently buzzed through our solar system was probably cosmic junk left over from some long-dead stellar civilisation. While very few other astrophysicists or astronomers agree with him, Loeb continues to argue that 'Oumuamua, also known as 1I/2017 U1, may well be humanity's first contact with an artifact of extraterrestrial intelligence; "considering an artificial origin, one possibility is that 'Oumuamua is a lightsail, floating in interstellar space as debris from advanced technological equipment" (Bialy & Loeb, 2018).

Loeb's book-length follow-up to his spectacular 2018 paper was his *New York Times* bestseller entitled *Extraterrestrial: The First Sign of Intelligent Life Beyond Earth*. For Avi Loeb, advanced extraterrestrial intelligence absolutely must exist, and on a vast scale, as thinking otherwise is just arrogant. "I start from the principle of modesty. You know, if we believe that we are alone and special

---

[3] For some, the above may already resonate, though the seventh and eighth chapters will provide a much deeper analysis of human "nature".

and unique, that shows arrogance", he told the AI researcher and podcaster Lex Fridman in an almost three-hour interview (2021). While this is not a strong logical basis for such belief, it has led Loeb to speculate along the Kardashev scale. We have not yet been deliberately contacted because we don't really matter. "I would think that we are sort of middle of the road, typical, forms of life, and that's why nobody pays attention to us. If you go down the street on the sidewalk and you see an ant you don't pay attention or special respect to that ant, you just continue to walk." We are not very interesting, not exciting, "so nobody cares about us". Because they have far surpassed us; "they could be a billion years old. And then imagine a billion-year technology. It would look like magic to us, you know, an approximation to God". And we must speak quietly to these gods, he told Fridman; "if you are inferior, there is a risk if you speak too loudly something bad may happen to you". Yet it doesn't matter how technologically advanced, or even staggeringly technologically advanced, such extraterrestrials might theoretically get, as psychologically and intellectually they can really be no more advanced than we are, albeit to fully convince you of this we will need to delve into the next two chapters, and consider first the role of reason, and then the possibilities offered by both machine intelligence and enhancing the biological with the technological.

Transition (Darwin's interpretation)	E.T. possible?	Dominion?	Spacefaring?
6. Evolution of multicellularity	✗	✗	✗
7. Eusocial colonies	? ✗ (Type I)	✓	✗
8. Dual inheritance system	✓ (Type II)	✓	✓

Because even if reason or enhancing the biological can take the process a little bit further, it does seem there can only ever be, at most, two patterns of extraterrestrial intelligence: maybe Type I and certainly Type II, the former a deeply brutal type that still accords to the billion-year evolutionary mould, and the latter a type that has managed, in one of only three argued ways, to change the billion-year mould. For Darwin, the patterns are a) potentially a deeply brutal one and b) the rootless outsider that only partially escapes its nature. For the group-selectionists and sociobiologists, the two patterns are a) the very common deeply brutal one and b) the very rare avuncular largesse one. Only one of the three answers can be correct – assuming we are treating group selection and sociobiology as different answers – but one of these three answers will be correct. So while we have cosmic dinosaurs everywhere, we will have, and can only ever have, one or two patterns of E.T. Everything beyond this is just tinkering. Anywhere in the universe, and across any epoch. Even for those who "could be a billion years old". Yet, *pace* Avi Loeb, given the inherent limitations to those two basic patterns, we actually matter very much indeed to the future of the universe, as we shall see.

5

REASON, AND THE RACE OF DEVILS PROBLEM

> "Yet across the gulf of space, minds that are to our minds as ours are to those of the beasts that perish, intellects vast and cool and unsympathetic, regarded this earth with envious eyes."
>
> – **H.G. Wells**, *The War of the Worlds*

The question for this chapter is to what extent reason can save us from what appear to be the severe limitations of Type I and Type II intelligence. As we have seen, evolution can happily furnish us with the unsympathetic, but can the natural world even create creatures with "intellects vast and cool", or with "intelligences greater than man's and yet as mortal as his own", as Wells imagined? Or "curious and dispassionate, observing us, as we would watch a bacterial culture in a dish of agar", as the astronomer Carl Sagan suggested (1980, p.308)? Extraterrestrials so psychologically and intellectually advanced that we might appear as Avi Loeb's "ants" to them, and they as his "gods" to us? Let us consider the race of devils problem.

The eighteenth-century philosopher Immanuel Kant wrote the following in his work *Perpetual Peace*. "The problem of organizing a state, however hard it may seem, can be solved even for a race of devils, if only they are intelligent. The problem is: 'Given a multitude of rational beings requiring universal laws for their preservation, but each of whom is secretly inclined to exempt himself from them, to establish a constitution in such a way that, although their private intentions conflict, they check each other, with the result that their public conduct is the same as if they had no such intentions'" (Arendt 1992, p. 17). A problem like this must be capable of solution, wrote Kant, and it does not require that we "know how to attain the moral improvement of men" but only that we should "know the mechanism of nature". Instead of genuine morality, he continued, the mechanism of nature brings morality to pass through selfish inclinations, which naturally conflict outwardly but which can be used by reason as a means for its own end, being the sovereignty of law, and for "promoting and securing internal and external peace" within the state.

Kant's race of devils – Volk von Teufeln – problem was a near-perfect anticipation of the theory of evolution by natural selection. There is an enormous literature detailing examples of "private intentions" conflicting, and of animals "secretly inclined to exempt" themselves from codes of behaviour that would benefit the group, and even within the most closely related insect colonies. We have looked at only a few examples, from Dawkins' blackheaded gulls swallowing a neighbour's unattended chick whole, to reproductive suppression in mole-rats, and queen-worker conflict in social insects. We have also looked at some of the underlying theories behind the above, from John Maynard Smith's evolutionarily stable strategies, to Hamilton's selfish herd theory across 2-dimensional space, and the "anti-social" mutations that Wynne-Edwards had denied would propagate. Nature is an arms race of attempts to cheat, countered by advances in the detection and prevention of cheating.

As mentioned, John Maynard Smith is the father of evolutionary game theory, and in nature it is not Kant's "reason as a means for its own end" that ends up creating the stability, it is iterative natural selection. Darwinian selection plays the role of the rational policy selector assumed in economic game theory. "Natural selection provides a dynamics which will, subject to constraints, cause a population to evolve towards an optimum" (Maynard Smith 1996, p.291). But we have also seen that non-inbred diploid inheritance cannot produce in vertebrates the necessary group sizes for organising a "state", or at least a large group. Loosely cohesive baboons can get up to groups of a few hundred, while the smarter and more closely co-operating chimps and bonobos live in fission-fusion societies of rarely above one hundred members.

While relative brain size in primates may be associated with factors including the cognitive complexity and cognitive flexibility needed to obtain a more transient diet (DeCasien *et al.*, 2017), and sexual selection may also have played a role, what higher capacities for reasoning can also conform to are the smaller group sizes necessary to counter more cunning cheating. As Dawkins once noted, "It is even possible that man's swollen brain, and his predisposition to reason mathematically, evolved as a mechanism of ever more devious cheating" (1989, p. 188). In single inheritance system nature, the vast numbers reflective of a "state" – the millions of co-operating driver and army ants for example – require transition seven colonial living, a high degree of hard-wiring, very close genetic relationships, and often key resource constraints.

So Kant's race of devils problem does indeed give us at least the template for our E.T. Type I, albeit with iterative natural selection replacing rationality. Instead of genuine morality, as Kant put it, it would be the selfish "mechanism of nature" that would organise this state, "promoting and securing internal and external peace". Of course it is still a contingent peace, the reproductive dictatorship of the naked mole-rat, a state controlled by coercion and violence, where "rival challengers for breeding status are usually attacked and killed", and where the social situation "changes dramatically when the queen's ... dominance begins to fail".

Originating in...	E.T.	Rationality?	Requires	Possible?
Race of Devils	Type I	Natural selection	Close relationships	Yes
Race of Devils	Type II	Cultural conditioning	Manipulation	Yes (Darwin)
Race of "Better Angels"	Type II?	Natural selection	Group selection	No (Darwin)
Race of Devils	Type III?	Natural selection	Perfect reason	No

And Kant's race of devils problem also gives us the initial template for E.T. Type II, because of course orthodox evolutionary theory says Type II begins life as a genetic (or equivalent) devil. But here it is neither "a multitude of rational beings" nor natural selection that provides the dynamics that permits an organised state. It is a second inheritance mechanism, it is the manipulation of cultural conditioning – religions that at the same time inspire enormous compassion and utterly murderous suicide bombings and holy wars, and humanistic philosophies that can simultaneously liberate coloured slaves while continuing to see them as physically different, as unequal, as "the position of superior and inferior" – that makes the state possible, and then makes possible the technological advances far beyond anything Type I appears capable of.

But here sociobiology (and group-selectionism) wants to replace Darwin's Type II with a different Type II, not a race of devils, but a race of, at the genetic level, "better angels". The Harvard sociobiologist Steven Pinker even wrote a 2011 bestselling book

– admired by everyone from Bill Gates ("the most inspiring book I've ever read") to Mark Zuckerberg – entitled *The Better Angels of Our Nature*. A title taken wholly without irony from the famous phrase by Abraham Lincoln, that lifelong white supremacist, at his inaugural address as US president where he confirmed that he had "no inclination" to "interfere with the institution of slavery". This was not Darwin's answer of the contingent better angels of manipulating culture, the compassionate religions, the empathetic humanistic philosophies; this was the sociobiological answer of the better angels of our genetic nature. This was not George Price's answer where culture "beat out" a billion-year malign nature; this was the sociobiological answer of culture gently holding the leash of our essentially benign genetic nature. This was Bob Trivers' answer that in human evolution "altruistic acts are dispensed freely", and Ed Wilson's answer that "morality evolved as instinct". Sure, Pinker wrote, culture could make us do cruel things, and post-Enlightenment culture and unfettered and unregulated free-market capitalism was making us do fewer cruel things now than mysticism and statism had previously made us do, but culture was manipulating a race of biologically better angels, manipulating the better angels of our nature, not manipulating Darwin's race of biological devils, not manipulating natural devils. Because humanity had, quite uniquely, out-evolved the billion-year genetic code. "Inspiring", enthused Bill Gates. Understandably, many desperately want the sociobiological (and group-selectionist) thesis to be true, an ur-thesis of a completely new direction, a turnabout direction, in biological evolution, and have always wanted it to be true, from the Christian Lord Kelvin and the anarchist Peter Kropotkin to Steven Pinker and Mark Zuckerberg, but all the evidence – mathematical, biological, logical and historical – is that this is nonsense. At least for Darwin and the gene-selectionists there is no, and there can be no, race of biological "better angels".

A PERFECTLY RATIONAL E.T.?

But now for the big question of the chapter. Within Darwin's model, can reason take natural creatures to what we shall provisionally call E.T. Type III? Admitting that we must start with nature's race of devils, but then following to its end-point Kant's prescription of a multitude of rational beings with "reason as a means for its own end"? John Maynard Smith considered the question a number of times over the years, and saw little problem with large-scale co-operation and cohesion if behaviour remains completely rational. Utterly self-interested rational devils would still realise that their long-term interests are best served by foregoing many of their short-term interests. Such devils would understand that seeking individual short-term gain by looking to take benefits without paying costs – a little bit of cheating here and there – will ultimately defeat long-term gain by reducing trust and therefore potential group size and organisational stability. Such cheating would ultimately lead to less trust and cohesion within the group that in consequence harms each individual.

As Maynard Smith realised, though, while maybe possible at the theoretical level, there are a number of problems with this idea at the practical level. The first problem is that evolution by natural selection is a process with no foresight, and works because it is. The efficient selection of purely random mutation. Natural selection generates small improvements, and thereby produces precise, but never perfect, contrivances. On the seemingly plausible assumption that Type III will require getting us to perfect, or at least near-perfect, rationality, perfection is something nature just does not do. Evolution is often a series of kludges and design tinkerings, as George Williams explained in his Science Masters' edition, *Plan & Purpose in Nature*. "We are in fact plagued with dysfunctional design features from head to toe, some resulting from evolutionary changes that may have been quite adaptive when they first occurred" (1996, p. 187). For example, he writes, all vertebrates, from fish to mammals, are capable of choking on food, due to a very early association between the respiratory and digestive systems. Thousands of people die each year from this ancient design bodge, and thousands more are only saved by the intervention of the Heimlich manoeuvre. The evolution of speech had to compromise a general mammalian evolutionary workaround that minimises the risk of choking, leaving us at worse risk than most other mammals.

The second problem, though, is why would natural selection ever need to get close to near-perfect rationality? Nature has no foresight, but it at least tends to bumble along to end-points that there is an evolutionary need for, and there seems to be no such evolutionary requirement here. Near-perfect rationality would effectively mimic the unworkable hypothesis (except in very small and specific cases) of genetic group selection. Getting to that point – that "mechanism of nature" that would organise this state, "promoting and securing internal and external peace" – would be very difficult, as it would be subverted at every step of the way, as until you have got to near-perfect rationality it can be undermined by cheating. If it requires, say, ten "steps" in the evolution of rationality to reach near-perfect forward-sighted ends-directed reason, it might or might not be stable once you get to step ten, but steps one through eight certainly seem to throw up a selective disadvantage in moving further towards ends-directed rationality. (And you may also need all to get to step ten simultaneously, as the group might otherwise be subvertable from the – by now – almost devilishly canny nines.) But the larger problem is what does step ten anyway offer that steps one through eight, and possibly even a much lower number than eight, do not? The large group size problem, such a benefit in some situations, has already been solved at the multicellular animal level by avoiding the need for close co-operation (think wildebeest), has been solved for close co-operation by very close genetic relationships and colonial living (think termites), and has been solved without close genetic relationships through a second inheritance mechanism and cultural manipulation (think humans). Reason is a tool of evolution, but near-perfect reason seems to be necessary for nothing other than solving the large group size problem, which has already been solved, and this alternative route would anyway be plagued with stability issues in getting to such an end-point.

The third problem is that larger brains not only do not prevent cheating, just promoting more intelligent cheating, but that the largest and most synaptically connected brain (as far as we know) that nature has so far managed to put together is not overly rational, and still seems *so very far* from the near-perfect rationality Type III would require. As John Maynard Smith put it, "the difficulty for any account of society that assumes that individuals behave rationally is partly that experimental psychologists find little support for such an optimistic view" (1992, p. 121). After describing the game theory computer simulations that have been run to demonstrate just what is possible if players continue to act rationally, Richard Dawkins relates that "sadly, however, when psychologists set up games of Iterated Prisoner's Dilemma between real humans, nearly all players succumb to envy, and therefore do relatively poorly in terms of money" (1989, p.220). It seems that many people, Dawkins continues, and "perhaps without even thinking about it", would rather "do down the other player than cooperate with the other player". Which is sort of what you might expect once you move from the ruthless logic of Type I to the contingency of Type II.

During the recent Covid and Brexit-related UK food and fuel shortages there was some interesting behaviour. There were no real fuel or food shortages until panics about possible fuel and food shortages took hold. At that stage people began hoarding not what they needed, but what they might need, creating the shortages. There is no argument here that they were not behaving "rationally" in an individual short-term sense, but their actions were undercutting others (particularly the most vulnerable) and the longer-term situation. A similar, but more extreme, version of the above behaviour can be seen in Moscow today. Ajay Goyal was founder of the influential 1998–2005 weekly newspaper the *Russia Journal*, and in his work *Uncovering Russia* he writes that: "Moscow drivers, for their part, seldom give way to ambulances or rescue squads" (2003, p.138). There is a very widespread belief in Russian cities that ambulance drivers are simply acting as private chauffeurs for wealthy Russians who want to get around quickly and avoid traffic jams. Muscovites have so lost faith in their civic government that many would prefer to block an ambulance that might be abusing its siren, even though if just a small portion of ambulances are still reacting to real emergencies you will be imperilling such victims, such victims that may one day even include you or your family.

The last forty years of cognitive psychology have been a growing realisation of just how poor our brains are at understanding the real world. The psychologist Daniel Kahneman won the Nobel Prize in economics by showing that the basis of modern economic theory had been wrong to think of agents as rational and with stable preferences. Kahneman believes the brain has two systems of mental operation. One is fast and depends heavily on intuition, but therefore often gets it wrong and can be easily fooled. The second is careful and slow, but lazy, and is forever having to make up for the errors of the faster system. Kahneman and his long-time colleague the late Amos Tversky once rigged a wheel of fortune to stop at one of two points, spun the wheel and asked people a totally unrelated question. The spin of an unconnected wheel of fortune should not have influenced the answers. But it did, and profoundly, and this is known in psychology as anchoring. As Kahneman notes, "there was no way to describe the anchoring effect of a wheel of fortune as reasonable" (2011, p.120). Kahneman has his critics within the field of bounded rationality first developed by Herbert A. Simon from the late 1940s, and most prominently the psychologist Gerd Gigerenzer at the Max Planck Institute for Human Development, but the fact that we are Type II evolutionary creatures might even help explain why the brain has both two systems of operation and bounded rationality. After all, we have two largely incompatible inheritance mechanisms, so who would ever have expected a single coherent mental system of operation? Type II cultural conditioning certainly helps explain Kahneman and others' larger insight that humans are so very often unreasonable and wildly self-deluding, up to and including philosophers and social scientists at Harvard.

HOW RATIONAL IS TYPE II, THEN?

If E.T. cannot be perfectly rational, how far might its rationality have nevertheless taken it? Let us consider the role of reason in the only Type II we have experience of.

> "If the great evolutionists still believe in the myth of free will, why should we trust them on anything?"
> – **astrophysicist** writing to **Will Provine**, 1 February 2008, after viewing biologists' responses to the *Cornell Evolution Project*

Free will, freedom of choice, freedom for any particular individual to have chosen differently, cannot exist in this or any possible universe. It is formally called the dilemma of determinism, given we inhabit a universe where all current and future events – at least from the point of view of human action – are necessitated by past ones. But it could as easily have been termed the dilemma of indeterminism, where indeterminism recognises events with no cause, as arguably (because this is still much debated) in the quantum world. As the philosopher Paul Russell puts it, "one horn of this dilemma is the argument that if an action was caused or necessitated, then it could not have been done freely, and hence the agent is not responsible for it. The other horn is the argument that if the action was not caused, then it is inexplicable and random, and thus it cannot be attributed to the agent, and hence, again, the agent cannot be responsible for it" (1995, p.14). As Russell spells out, the dilemma of determinism has stark implications because if our actions are caused, then we cannot be responsible for them, but if they are not caused, we cannot be responsible for them. "Whether we affirm or deny necessity and determinism, it is impossible to make any coherent sense of moral freedom and responsibility."

We *can* arguably debate around the concept of responsibility, but we *cannot* reasonably debate around the concept of moral freedom. The absence of freedom to have chosen differently is woven into the very fabric of the cosmos; it is built into the logic of the universe. And in a world without free will "luck swallows everything", as the Oxford philosopher Galen Strawson puts it (1998), and all of life is reduced to the pure lottery of biological and environmental luck. Before I was mentored in evolutionary paradoxes by George Williams my background was philosophy and metaphysics. Philosophers like my colleagues Bruce Waller (1990, 2006), Derk Pereboom (2001, 2007) and Richard Double (1990, 2002) have written extensively on why, logically, luck swallows everything in human life. The reason that the publishing sensation Yuval Noah Harari can assert as self-evident that "humans have no free will" (2018, p.300), and that this is "such a radical message" with "sinister implications" (p.251), is largely because of Bruce's, Derk's and Richard's earlier, and courageously contrarian, analyses. And the above leads to some fascinating insights. For 3,000 years the most sought-after principle in ethics has been the possibility of an objective base to knowledge. The absence of free choice – and because it is that extraordinarily rare beast, a nontrivial question that can be reduced to pure logic alone – is probably the *only* ethical absolute that humankind can or will *ever* have access to. Just think about that for a moment. "Luck swallows everything" is the single and solitary moral fact that is built into the workings of the universe. The only moral fact we can ever know, a moral fact built into the structure of creation, yet … at least 85% (and probably closer to 95%) of the Western public is wholly ignorant of, and currently incapable of even reasoning towards, this solitary absolute ethical element (Bamfield & Horton, 2009; Allen & Dimock, 2007).

Now of course there are explanations for why a Type II intellect might raise their personal intuitions and deeply-held beliefs over logic and reason. Some thinkers, such as Spinoza, suggested that humans mistakenly believe we have freedom of choice because we are not conscious of the determined nature of our desires, arguing in effect that we don't know our own minds. And, indeed, the brain receives over ten million bits of information each second from various sensory systems, but the conscious mind can handle only up to about fifty bits per second (Mlodinow, 2012). The unconscious does the heavy lifting, and consciousness seems to be a deliberately simplified model of the world to stop us drowning in data. In contrast David Hume suggested that it was the experience of indecision that made us mistakenly think we have free will, while more recently Seth Lloyd, a physicist and mathematician at the Santa Fe Institute, has suggested (2012) that it is perhaps the intrinsic computational unpredictability of the decision-making process that gives us the impression that we are freely choosing. Whatever the ultimate explanation or explanations, as with Kahneman above within the last forty years of cognitive psychology there has been a growing realisation of just how inadequate our brains are at understanding the real world, and just how poor we are at elevating the limited capacity for reasoning we even start with. And cognitive theorists are only too aware that our minds play tricks on us: we invent reality within our minds, with our brains making up all the missing pieces. Neuroscientists call the explanations our minds create after an event confabulations; a form of "honest lying", because the individual is unaware that the information is false. Experiments show that eyewitnesses remember far less than they themselves insist upon, because many of their recollections have been filled in or invented by their brains. And neuroscience is showing the very feeling of intention to be an illusion created by the brain. The most famous experiments throwing doubt on classical conceptions of free will were performed by Benjamin Libet in the early 1980s. Libet got volunteers to watch a clock, move their wrists, and report at what time on the clock they chose to move. At the same time Libet measured electrical activity over their brains, and found that the neural preparation to move preceded the volunteers' conscious awareness of the intention to move by over 300 milliseconds. So there was a spike of brain activity 0.3 seconds before the volunteers chose to move their wrists, or to put it another way the volunteers' brains prepared to move before the volunteers consciously decided to move. Libet's work has now been verified, modernised and extended by numerous other neuroscientists such as John-Dylan Haynes, Itzhak Fried and Patrick Haggard, while Haynes co-authored a 2008 study issued by the Max Planck Institute which showed that entire seconds before we are aware of making a decision our brains have already made the decision.

But the above does not *excuse* Type II belief in freedom of choice, even if it *explains* it. Logic has always ruled out the possibility of free choice even before late twentieth-century neuroscience got in on the act, and Charles Darwin called freedom of choice a "general delusion" (Barrett *et al.* 1987, p.608), while Einstein could "not at all" be persuaded by the conceit (1954, p.8). As mentioned earlier, Friedrich Nietzsche called free choice "a kind of logical rape and unnaturalness" (1886, p.50). So perhaps it does not matter that the general public is not particularly rational, so long as our top thinkers, our philosophers and our scientists are at least reliably rational and open-minded. Surely philosophers and scientists can then educate, guide and inform the less rational public. But it appears that most scientists and philosophers cannot be trusted to tell the truth, cannot be trusted to educate, at least not when it really matters. A decade and a half ago Greg Graffin, a graduate student of my late friend the Cornell biologist Will Provine, contacted 272 leading biologists and asked for their responses to a number of metaphysical questions. It was originally called the *Cornell Evolution Project, a Survey of Evolution and Religious Belief* (see www.polypterus.org), and Graffin's list of 272 reads like a *Who's Who* of the great and the good of the last half-century of evolutionary biology, from Francis Crick, John Maynard Smith – misfiled under "Smith, J.M." – and George C. Williams, to Lynn Margulis, Richard Dawkins, and Ed Wilson. Provine told me that Graffin discovered that while fewer than 5% of these scientists expressed belief in the existence of a creator god, fully 80% expressed belief in the existence of free will. So what is going on here?

"Free will is not a scientific concept: it means 'not caused by anything,' and the scientific worldview can only seek causes", said the Harvard sociobiologist Steven Pinker in interview (Blume 1998, p.155). Yet at the same time Pinker has said that he believes "that science and ethics are two self-contained systems played out among the same entities in the world, just as poker and bridge are different games played with the same fifty-two-card deck" (1997, p.55). What Pinker seems to mean here is that scientists have no duty to tell the truth when such truths risk undermining economic and political realities, and that indeed for Pinker they seem to have a duty not to tell the truth in such situations. The *Guardian* journalist Alex Blasdel spent days interviewing Pinker in summer 2021, and while his long article was certainly no unfriendly hatchet job (at one point he ended up "wearing an old pair of Pinker's shorts"), it did mention some uncomfortable history, including Pinker's eagerness to plunge into "acrimonious debates over gender, race and progress" (Blasdel, 2021). Pinker has been called "the world's most prominent defender of the status quo", and he is on first-name terms with a number of the world's richest and most powerful people. Mark Zuckerberg has included Pinker's books on a list of what to read at Davos, and Pinker apparently keeps a directory on his phone of the two dozen or so heads of state, royalty and other leaders who have asked him for an audience. His friend Bill Gates ("Bill's got a pretty nimble mind") has identified Pinker as the part of the "conservative centre" in interview with the *New York Times*, and Pinker is the stand-out guru to a faction known as the New Optimists, a largely ideological grouping that militates for the power of unfettered and unregulated free markets in driving human history. But it is unfair to single out the conservative centre sociobiologist Steven Pinker as unwilling to challenge widely believed myths that help maintain the social and economic status quo. The somewhat more left-leaning Richard Dawkins has also called the myth of free will a "useful fiction", one that is conveniently "short-cutting a truer analysis of what is going on in the world in which we have to live" (quoted in Dennett 2008, p.253). The one time Dawkins tried to highlight that the concept of free will was an intellectual fraud, for an *Edge.org* survey of the world's "most dangerous ideas" – a survey intended to draw a direct parallel with other scientific "discoveries that were considered socially, morally, or emotionally dangerous in their time" – he received an icy reception from his friend the philosopher Daniel C. Dennett. Dennett went on to reveal his satisfaction that Dawkins "later regretted sending and tried … to retract" (p.253) his *Edge.org* article.

There are biologists who truly believe, or at least have believed, in the notion of freedom to choose, even though such an idea has always been wholly illogical. The sociobiologist Ed Wilson believed in the possibility ("the paradox of determinism and free will …

might even be reduced in status to an empirical problem in physics and biology", 1978, p.77), as did the Nobelist Gerald Edelman ("a human being has a degree of free will", 1992, p.170). But on closer analysis it appears the reason the *Cornell Evolution Project* uncovered so many leading biologists professing belief in free will was not because most genuinely believed in it, but because most biologists are not at all averse to "short-cutting a truer analysis of what is going on". Though it would be unfair to leave the impression that biologists are the only scientists to trade in useful fictions, because it does appear physicists have often acted just as questionably. Indeed, physicists appear more likely to really believe in the illogic of free choice, and it is probably not unrelated that physics has a history of attracting very senior mathematicians with a weakness for mysticism. In the *European Journal of Physics* Juan Marin (2009) catalogued the legends of quantum mechanics who had such a failing, and this is sometimes even referred to as quantum mysticism. Marin mentioned Heisenberg and Schrodinger, and particularly Wolfgang Pauli. Pauli was a close collaborator with, and shared the same anti-rationalism as, the psychoanalyst, occultist, alchemist, and sometime parapsychologist Carl Jung. In a further piece of research Marin has shown that Hermann Weyl was immersed in mysticism, that David Bohm wrote extensively on Eastern mysticism, and that Max Planck was a propagandist for Western religious mysticism (see Zyga, 2009). However, so far we are really only scratching the surface of what is going wrong with Type II rationality.

DOES RATIONALITY EVEN MATTER IF WE CANNOT BE TRUSTED TO TELL THE TRUTH?

In a fairly recent survey of 2,000 top philosophers around the globe, more than 86% recognised freedom of choice to be a logical impossibility. That means that almost 14% of philosophers still believe in free choice, about half "fully accepting" and half "leaning toward", but this group is "dominated by theism, a rejection of naturalism … and non-physicalism about the mind" (Bourget & Chalmers 2014, p.489). Yet even though over 86% of philosophers know free choice to be a myth, the vast majority of this number do not want to disabuse the public over the myth. Dan Dennett has written that "when we consider whether free will is an illusion or reality, we are looking into an abyss" (2008, p.249), while another leading free will apologist philosopher, Saul Smilansky, has called you non-philosophers "fragile plants" who "need to be defended from the chill of the ultimate perspective in the hothouse of illusion" (2011, p.436). At the same time Smilansky has argued that free will is a fraud that academics can freely and legitimately perpetrate on Western publics, in part because he believes "the threat of political manipulation and the like is less acute here" (2000, p.271).

What we are seeing at this point is partly woefully misplaced intellectual condescension and arrogance, but also the emergence of something more worrying. An admission that thinkers and intellectuals don't have a duty to tell the truth, at least not when it really matters and they don't fancy doing so. Philosophers shouldn't mislead the public over the small things in life, this argument seems to go, but they – or at least not much south of 86% of them – reserve the right to deceive and misrepresent when they, and they alone, consider it socially and economically necessary. The only provable moral fact that humanity can probably ever know, and we cannot trust around 90% of our smartest thinkers to even tell the truth about it. Which is surely sort of getting close to game over for the conceit of humankind as the rational ape, or *H. sapiens* the "wise" man, as Linnaeus named us. The wise animal, as Linnaeus was effectively naming all natural Type IIs, on this planet or anywhere. Because does contemplating rationality even matter if our smartest thinkers cannot be trusted to tell the truth, and if, as Harvard's Steven Pinker advised previously, ethics and scientific knowledge should be treated as two separate "games"? We should trust philosophers and scientists, except when it really matters, because then they very well may not be telling the truth. Which leaves us in the worst of all possible worlds of never quite knowing when we can trust them, and so "why should we trust them on anything?", as the astrophysicist writing to Will Provine legitimately enquired. For now we will have no idea when not to trust these thinkers, unless we already have an insight into the knowledge and economic systems they fear. They wish to be able to pick and choose which truths you "fragile plants" genuinely need to know. As Dennett puts it, the public eventually realising there is no possible freedom of choice is a thought he finds "almost too grim to contemplate" (1984, p.168).

Yet it is not just that philosophers like Dennett are openly trying to convince the public to believe in something they themselves do not believe in, because at least some of the time they appear to be either trying to convince themselves, or to be deeply confused on the issue, another reason why Type II rationality and wisdom seems to be remarkably weak tea. Dennett is a Darwinian, a naturalist, a materialist, fully aware that freedom of choice is logically impossible, fully aware that luck swallows everything, but this doesn't stop him performing intellectual u-turns at the drop of a hat. Dennett writes erroneously that when we are making important decisions we undertake an internal monologue or debate but at some point we stop the deliberation "in the full knowledge that I could have considered further" (1978, p.297). Yet as my colleague the anti-free will philosopher Bruce Waller notes, this "glib suggestion" that everyone, no matter what their capacities and resources, could have carried out the same level of self-deliberation, internal monologue and self-improvement "is both shallow and false" (2011, p.163). In his long free will apologist tome *Elbow*

Room, Dennett has claimed that "there is elbow room for skill in between lucky success and unlucky failure" (1984, p.97). But it is utterly disingenuous for a Darwinian materialist to state that there is elbow room for skill between good luck and poor luck, which is exactly equivalent to arguing that $1 + 1 = 3$. According to Dennett there is something left over when one takes away the lottery of biology and environment. You can have good or bad fortune in internal causes (biology), and you can have good or bad fortune in external causes (environment) but, for Dennett, there seems to be something beyond normal causation, some form of uncaused skill, or phantom elbow.

"I must look like a spin doctor to those who are unconvinced by my carefully marshalled analyses", Dennett writes (2012b, p.20), and hence we must just assume the existence of free choice "within limits we take care not to examine too closely" (1984, p.164). But it is not just Dennett who is a materialist trying to sneak mystical self-creation and ghost elbows in through the back door, or by turning a blind eye to logic. Here are some other influential philosophical examples. Jonathan Jacobs writes that "the inability of the ethically disabled agent to overcome that condition is not exclusively a matter of bad constitutive luck" (2001, p.81). But unless you are one of the less than 14% of philosophers who still believe in self-creation, which Jacobs is not supposed to be, it is exclusively a matter of bad constitutive luck. Or there is Charles Taylor erroneously writing that "self-resolution in a strong sense ... is within limits always up to us" (1985, p.42). Or Gary Watson writing that "the force of the example does not depend on a belief in the *inevitability* of the upshot. ... The thought is not 'It had to be!'" (2004, p.243), when in fact the logical thought is only of inevitability and that "It had to be!". Waller (2011, pp.115–31) provides a whistle-blower list of other examples of supposed philosophical naturalists and materialists ultimately espousing a form of voodoo self-creation.

We are spending so long on the free will debates precisely because this goes to the very heart of human – and thus all Type II – rationality, or lack thereof, while at the same time allowing us to paint across a unique canvass there should logically be no debate over, being knowledge built into the fabric of the universe. For the key to Type II behaviour, the key to rising above the indifference of Type I existence, seems to be not rationality; it is in a very real sense the Type II *lack* of rationality, and this should become a revolution in the way we understand the human animal, and the human mind. Type II overwriting seems to work *because* we are susceptible to these somewhat irrational prejudices, hopes and fears; largely predisposed to something no perfectly rational creature would be, could be, susceptible to. Irrationality and dogmatic unreason are under this viewpoint not occasional and ephemeral by-products of the human condition; they are in an important, albeit somewhat tragic, sense fundamental to what it means to be human, fundamental to becoming human in the first place. We should by this point of the book expect our smartest thinkers to be not wholly rational; storm-tossed outsiders, complex amalgams of other people, highly educated but all too often still festering pools of private school privilege, spin, and condescension; "their lives a mimicry, their passions a quotation", as Wilde put it earlier.

We must now mention that once philosophers go into detail on free will, the arguments go beyond the intellectually embarrassing and touch upon the truly shameful. As Bruce Waller has noted, Dennett does not even want a social and moral system that is fair, rather he has written that the system should be "fair *enough*", and that Dennett then defines *fair enough* as the interests of the lucky. "Dennett seems comfortable with 'fair enough', and he can champion such a system and not blink" (Waller, 2012). Dennett responded to Waller that he "dismisses as absurd my claim that it is fair because luck averages out in the long run. ... The luck averaged out in ninety-nine percent of the population. ... If anything like that were true, my claim would not be absurd at all" (2012; see also 2012a). For Dennett, the system is "fair enough" because it worked out for the vast majority. The fact that it was manifestly unfair for a small minority ("but *only* 3 million!" emphasis is his) doesn't even seem to impact on Dennett's consideration. By the way, Dennett appears to be using the above example to suggest that 99% of poor black Americans get exactly the same overall opportunities as 99% of rich white Americans. Really? And Dennett then goes on to say we anyway have a right to be unjust towards the minority, because "we are rewarded for adopting this strategy" by the greater number going on to display responsible behaviour (1984, p.164). Is this not the most extraordinary and smug assertion – that the majority learn responsibility and fairness through deception and the practice of injustice towards the minority ("but *only* 3 million!")?

Yet Dennett's writing is not even the most extreme in the literature, or particularly unusual; in fact, it is commonplace. Another tradition of liberal – and largely North American – free will apologist philosophy, termed essentialism, goes further in seeming to claim that the least fortunate are, biologically, often not our equals, and are thus deserving of lesser consideration. Gary Watson, mentioned above, is one of the leaders of the essentialist tradition, and Watson tells us that he has rejected that "troubling ... sense of equality with the other" (2004, p.245), denied the "ontological shudder" that would come with thinking we might share the same essential biological nature as them. Skin colour and nose shape do not drive this ghettoisation, but biology still does. To contrast the logic of these two leading schools, Dennett attempts to deny that luck swallows everything by claiming that unfairness to some does not matter, and that anyway we all, white or black, rich or poor, get the same opportunities in life. Watson changes Dennett's argument such that the unlucky no longer count within the fairness estimation. Dennett tried to show the scales balanced by in effect keeping a thumb on one side of the mechanism;

Watson just removes multiple people from one side of the moral luck calculation until it does start to balance out.

We are trying to get a handle on the extent of Type II logic, both here and on any other planet, so please understand that Dennett and Watson effectively represent over 75% of philosophers. There is also that 8 to 14% that either fully believes in freedom of choice or just that it may somehow be possible, a tradition today perhaps most associated with the American philosopher John Searle, but there are extensive writings pulling this tradition to pieces.[4] All together, more than 90% of leading philosophers are trying to deny what must surely be the apotheosis of ethical knowledge, that luck swallows everything in human life. Now there are of course reasons why the vast majority of philosophers dissemble over free will. Bruce Waller has wryly suggested that most philosophers "hold privileged and very comfortable positions in society, and we like to think we justly deserve those special benefits; we've accomplished much, and we are delighted to claim credit for it" (2011, p.307). It should also be recognised that Dan Dennett happens to be Daniel Clement Dennett III, whose father, presumably Daniel C. Dennett II, was a diplomat, and sent his privileged son to Harvard. And because philosophy is a non-vocational subject with no guaranteed utility in the workforce, the great majority of Anglo-American philosophers come from similar economically and socially advantaged backgrounds where you do not need to worry about repaying student debt, and this almost always affects their reasoning. Yet note that Dennett, Watson, Searle and Smilansky perhaps deserve more credit than most philosophers, because at least they are willing to put pen to paper. Most of that 90% of

---

4   He had not tried to solve the problem of free will, "Searle wrote, after nineteen pages of trying to solve the problem of free will. The flaw with each of his arguments, Searle admitted, is 'to see how the consciousness of the system could give it a causal efficacy that is not deterministic' (2000, p.21). In other words, we are back to the problem of determinism ruling out free choice" (Miles 2015, pp.17–18).

philosophers ride upon Dennett, Watson, Searle and Smilansky's coattails, and wish to be *de facto* spin doctors without actually raising their heads above the parapet and making the necessary detailed attempts to try to seek – inevitably unsuccessful – ways around the "luck swallows everything" moral luck abyss built into the workings of the cosmos.

We need there to be a robust sense of rationality, plus for reason to be an objective end in itself, if Type IIs are to have any hope of escaping both the indifference of their first inheritance system and the contingency of their second inheritance system – and notwithstanding the encroaching paradox we have touched on regarding *irrationality* appearing to be the very key to Type II overwriting, where Type IIs seemingly can't afford to be overly rational – yet consider what we actually see, because it doesn't get more absurd than this. The one, single, ethical fact we can prove, yet instead of using it as the objective basis of human morality, it ends up suggesting to us that 90% of our most logical thinkers shouldn't be trusted with a pair of sharp scissors. Almost wherever we look, we humans, we Type II intellects, are blind to truth and knowledge, and largely *because* of the very psychology that frees us from Type I indifference in the first place. You cannot trust humanist academics, and of course you cannot trust the general public either. Just consider religion for a moment; the word *Islam* is derived from the Arabic root *istaslama* which means submission (to God), surrender, giving in, and obedience without question, while the word *Muslim* come from the same root and means "one who surrenders". But intellectual surrender and obedience without question is anathema to reason, anathema to the only force in a universe without free will that might just have been able to raise a Type II intellect above the contingency of its second inheritance system.

Human psychology freed us from evolution's programme of amoral indifference and allowed us to build everything that we see round us, and everything that matters to us, from civilisation and peace to love and prosperity, so we cannot bemoan Type II overwriting, or the psychology that underlies it. But it does mean that in a very real sense unpredictable irrationality is absolutely integral to the possibility of Type II existence, here or anywhere in the universe. Or at least for biological deliberators, human or extraterrestrial, because as we shall see in the next chapter the rules are going to be somewhat different for mechanical deliberators. And even scarier.

6

TRANSHUMANISM, PLUS THE EXISTENTIAL THREAT OF AI

> "Success in creating AI would be the biggest event in human history. Unfortunately, it might also be the last."
>
> – **Stephen Hawking**, astrophysicist (Hawking *et al.*, 2014)

There is currently much thought and discussion over artificial, or machine, intelligence, AI, and transhumanism including augmented intelligence – humans, and perhaps extraterrestrials, who may use technology to enhance what nature provided – but there has been no serious attempt to reconcile such thinking to orthodox Darwinian theory, or at least Darwin's key eighth transition. An attempt must therefore be made to understand whether artificial intelligence or enhanced consciousness can take us beyond Types I and II, being the only possible patterns for deliberative life the natural world can give us. And the implications do appear to be profound, and again to be profound under all three extant evolutionary traditions.

	Inheritance system	Form	Intelligence level	Character
Machine intelligence	Single	Type I	Type II+	Extermination
Transhumanist intelligence	Dual	Type II (or → Type I)	Type II (or → Type I+)	Contingent (or → Type I)

MACHINE INTELLIGENCE

First some background, and while I do not pretend to be an authority on AI, this part of the book is vital, as it does seem the machine intelligence community has been suffering under serious misapprehensions that put us all at significant risk. Many modern approaches to developing advanced machine intelligence are conspicuously evolutionary, so it may well be evolution, and not computer science, that in the long run will define the character of AI. Traditional AI grew out of the work of the codebreakers in World War Two who ended up designing vast machines to crunch information. After the war this led early AI practitioners to pursue a path of trying to simulate human intelligence through symbolic representation of problems, and by having algorithms follow an index of set rules, and examples were the chess supercomputers in the 1990s designed to take on and beat the greatest grandmasters using what is termed brute force artificial intelligence. Some of the limitations with such expert systems included their handling of certainty plus the growing need for immense lists of pre-programmed rules. In contrast non-symbolic deep learning, or pattern recognition where data must pass through more than one – the "deep" in deep learning – optimising and recombining node layer between input and output, has been driving the more recent decade-long boom in AI, and uses artificial neural networks (ANNs) for everything from virtual assistants to self-driving cars. Across 2017–18 the AlphaZero algorithm played chess so well that human grandmasters concluded that for the first time a machine truly understood chess, as the program apparently played with real insight and even very un-machine-like risk-taking. Here data passes through webs of mathematics loosely modelled on how information passes through brain cells, where connections between parts of the network adjust as data is processed and decisions are updated, a self-corrective feedback loop leading to flexibility in the handling of future data. ANNs are part of an approach to AI termed artificial life, ALife or A-Life, and the first ALife conference was held in 1987. ALife can go beyond just mimicking the neurones within animal brains and can mimic the evolutionary processes that give rise to animal brains. Neuroevolution is AI that employs evolutionary algorithms, being mechanisms inspired by biological evolution, such as mutation and selection, to create ANNs, and evolutionary robotics is an arena where robots equipped with ANNs progress under artificial evolution. Machines are evolved by a process that imitates natural, or at least artificial, selection, where inchoate robots are allowed to compete, with the fittest being used to breed the next generation of robots, and so on. The artificial neural network route had largely fallen out of favour in the '70s and early '80s as there were arguments that such networks could never be sufficiently powerful, and though neural networks re-emerged in the late 1980s they subsided again in the early 2000s, only to be the route that was then vindicated in 2012 by a series of experiments successfully handling large levels of data.

In recent years most of the focus in AI has been about supersizing, and devoting increasing computing power to any problem, though this worries some that scale-up AI is becoming more and more the preserve of giant corporations with vast programming and financial resources, and worries others that continuous scaling is producing diminishing, and maybe intellectually finite, returns. A recent product, Megatron, mimicking human language and developed by Microsoft and chip manufacturer Nvidia, required more than a month of supercomputer access, and training datasets with 530 billion parameters. This is guidance in pattern recognition; show a neural network enough images of dogs versus cats, and reward and reinforce the optimising algorithm when it starts to notice the important differences, and eventually it becomes pretty good at differentiating, even when presented with a new breed. However, deep learning can still throw up crass or toxic mistakes, making it currently inappropriate for the highest stakes fields, and particularly when encountering outliers significantly different from the datasets on which it was originally trained. Deep-learning AIs lack comprehension, and even the best language AIs have been called "stochastic parrots" (Bender *et al.*, 2021), repeating back to humans variations on what we initially feed in, including our biases, in little beyond a semblance of understanding. Though that doesn't mean that people don't continue to be suckered by the responses of these pattern-matching chatbots (see Metz, 2022). The Santa Fe Institute's Melanie Mitchell comments that "like all AI systems of the past, deep-learning systems can exhibit brittleness – unpredictable errors when facing situations that differ from the training data", as these algorithms don't learn concepts, they learn "statistical associations … they learn shortcuts", as compounded by the problem that their "decision-making mechanisms can be quite opaque" (2021).

The above is simultaneously termed narrow (as opposed to general) AI, weak (as opposed to strong) AI, or simply machine learning, ML, and containing the paradigms of supervised learning (labelled datasets), unsupervised learning (working with the much larger group of raw unstructured datasets), and reinforcement learning. The holy grail for the machine intelligence field, however, is still artificial general intelligence (AGI), often simply called full, true, or human-level AI; high-level reasoning and general intelligent action including potentially consciousness, though due to the additional technical and conceptual complexity and the longer payback time horizon of AGI, being multiple decades not years or perhaps even months, most current focus remains on narrow and weak AI. Nevertheless, a 2020 study by the Global Catastrophic Risk Institute identified seventy-two active strong AI R&D projects spread across thirty-seven countries (Fitzgerald et al., 2020). The GCRI noted that about half of the projects are with private corporations, with academic institutions as the next most common establishment type. Nine projects had identifiable military connections, though mostly basic research. Almost half of the projects are based in the US, while the dangers may be magnified where one considers the authoritarian states that are both disregarding of international standards of behaviour and tightly control their newsfeeds. There are five AGI projects based in China. There are three AGI projects being led from a viciously aggressive, and highly corrupt, Russia, and Vladimir Putin told schoolchildren in 2017 that whoever becomes the leader in AI "will become the ruler of the world", according to one Russian state-controlled television network (RT, 2017). The GCRI commented that "most projects are not active on AGI safety issues, and some are openly dismissive of AGI safety concerns" (p.3), while stressing that others do have a significant emphasis on safety.

Some high-profile commentators, including Elon Musk, Bill Gates and the late Stephen Hawking, have tried to highlight that AI may present a sea change in technological threats, but their warnings have often been laughed off. The Harvard cognitive scientist Steven Pinker, in an essay entitled *Thinking Does Not Imply Subjugating*, has written that "AI dystopias project a parochial alpha-male psychology onto the concept of intelligence", with dystopians missing the point that AI may "develop along female lines" with "no desire to annihilate innocents or dominate" (2015). This is a somewhat bizarre reinvention of the last half-century of evolutionary orthodoxy, if you recall Amy Parish's comments that captive female bonobos establish dominance over males by overt aggression (Parish compiled records of injuries due to fighting and "found that all the wounds were caused when females, often collectively, attacked males" – De Waal 1997, p.189), or that in ground squirrels it is the female that "may raid the nest of a competitor and kill all the young", or that the foreign policy aim of those female-dominated ant societies is "territorial conquest, and genocidal annihilation of neighboring colonies whenever possible". Pinker's complacency stems partly from his sociobiological belief that Darwin was fundamentally wrong about both the evolutionary transition to intelligence and human biological development, but also from his personal political faith that technological progress and reason go hand in hand.

But such complacency – and lack of awareness of evolutionary paradigms – is also entrenched within large parts of Silicon Valley. Responding to Musk's and Hawking's concerns, Eric Schmidt, CEO and then executive chairman of Google in the decade and a half to 2015, and subsequently executive chair of Google's parent Alphabet until 2017, told a panel in Stockholm that such fears were misplaced. Schmidt noted that "neither is a computer scientist. Hawking is a physicist and Musk is an engineer", and that if there were to be any sort of danger "don't you think humans would then go about turning these computers off?" (Tung, 2016). Musk had previously suggested in interview that of all the companies investing in AI only one worried him. "Musk repeated his answer, suggesting his eye was on Google. 'There's only one'" (McCormick, 2016). As of 2020, two of the four largest AGI projects identified by the GCRI were Google's DeepMind and OpenAI, and where in 2019 Microsoft invested US$1 billion in OpenAI "explicitly to develop AGI by eventually scaling Microsoft Azure capabilities", though it should be mentioned that the GCRI categorised both DeepMind and OpenAI as projects "engaged on safety". Maybe Schmidt is right that Hawking was "only" a physicist, and Musk is "only" an engineer. But Darwin and George Williams were "only" evolutionary biologists, yet they understood the paradoxes behind the evolution of intelligence, natural or artificial, far better than anyone currently working at Microsoft or Google.

STRONG AI, THE SINGULARITY, AND DARWIN

A 2015 conference for AI experts in Puerto Rico found that on average most thought human-level artificial intelligence would be reached within forty years, although strong AI breakthroughs have been repeatedly over-hyped, and some at the conference thought it would still take centuries. Quoting Nick Bostrom from back in 2005, *New York Times* bestselling author on AI and the founding director of Oxford University's Future of Humanity Institute, what matters from a practical perspective "is whether and, if so, *when*" computers will be able to match human performance in tasks involving general reasoning ability. "With the benefit of hindsight, we can say that many of the early AI researchers turned out to be overoptimistic about the timescale for this hypothetical development." Bostrom went on to add that the fact that we have not yet reached human-level artificial intelligence does not mean that we never will, and he noted a number of computational or robotics experts, including Hans Moravec, Ray Kurzweil, and the late Marvin Minsky, who had put forward reasons for taking seriously the possibility "that this could happen within the first half of this century".

Yet given the hypothesised capacity for computers able to match human performance in tasks involving general reasoning ability, be that within the first half of this century, the second half, or still hundreds of years away, almost all within the AI field have an opinion on what is called the "singularity". The technological singularity is the day when computing power supposedly becomes so potent and continually self-enhancing, recursively self-improving, that it far surpasses the merely human, leading to artificial superintelligence. The term comes from a 1958 meeting between the computer scientist John von Neumann and the Polish mathematician Stanislaw Ulam. Bostrom records Ulam's recollection: "One conversation centered on the ever accelerating progress of technology and changes in the mode of human life, which gives the appearance of approaching some essential singularity in the history of the race beyond which human affairs, as we know them, could not continue". Bostrom noted that while von Neumann spoke of a discontinuity, nowadays the concept of the singularity often refers to a more specific prediction: "namely, that the creation of self-improving artificial intelligence will at some point result in radical changes within a very short time span", and as first clearly stated in 1965 by the statistician I.J. Good: "an ultraintelligent machine could design even better machines; there would then unquestionably be an 'intelligence explosion,' and the intelligence of man would be left far behind".

Fear, or enthusiastic anticipation, of the singularity dominates large parts of the artificial intelligence and techno-futurism communities, though not all share precisely the same concerns. A July 2015 open letter seeking to prevent the use of autonomous AI in warfare was signed by scientists and technologists including Elon Musk, Apple co-founder Steve Wozniak, and Stephen Hawking, and hosted on the website of the Future of Life Institute. The FLI is at the forefront of policing the risks and rewards of both weak and strong AI, and the following year spelled out in more detail the risks posed by machine intelligence. The FLI concluded that the fear of machines turning evil is a red herring, because the real worry isn't malevolence, but competence. "A superintelligent AI is by definition very good at attaining its goals, whatever they may be, so we need to ensure that its goals are aligned with ours. Humans don't generally hate ants, but we're more intelligent than they are – so if we want to build a hydroelectric dam and there's an anthill there, too bad for the ants. The beneficial-AI movement wants to avoid placing humanity in the position of those ants" (2016).

It seems we cannot escape being compared to ants, either by Avi Loeb's extraterrestrials, or the FLI's superintelligent machines. So, with this background, what can Darwin bring to the conversation? The first consideration must be that a pure machine intelligence, an evolving machine intelligence effectively allowed to roam free intellectually, is being programmed through a single inheritance mechanism, a single programming system. *And the only experience we have in a billion years of evolution, across a billion different species templates, and across the entire mathematical logic of natural selection, is that if you programme an evolving system through one inheritance mechanism you end up with "no evil and no good, nothing but blind, pitiless indifference".* It sort of no longer matters if people with vastly more knowledge of AI than the author have tried to pooh-pooh this concern, because this is the only experience we have to date to go on. And experience must trump disciplinary antipathy here. This consideration, emanating first from Darwin's own reasoning, and then later from the mathematics of the gene-selectionists, has to be allowed for, has to be confronted, and not swept under the carpet. Particularly not by those who have failed to recognise the transitions-through-seven behavioural and intellectual paradox that is being presented under each of the three extant evolutionary traditions.

Now it may be that no machine will ever have "roamed free" intellectually. It may be that no machine intelligence can ever be "pure" in this sense of iterative evolution through a single inheritance mechanism. We have already spoken about "supervised" ML and "reinforcement" machine learning. Every artificial Type I intelligence ever built – and I guess that is what they must be characterised as, as a first supposition – has been put together, at least initially, by a Type II intellect. Some AI guru may one day be able to convince us that this therefore makes machine learning, and all other pathways to machine deliberation, Type II intelligence, or at least non-Type I intelligence. That in effect we are already unconsciously inserting artificial sub-routines to mirror how a second inheritance mechanism can act, some possibly serendipitous analogue to the science fiction writer Isaac Asimov's deliberately civilising "three laws of robotics", albeit we should need a lot of convincing. And the reason we should require a heck of a lot of convincing is tied up in that table from the earlier chapters. This table is the reason the galaxy is almost certainly teeming with giant dinosaur-like creatures, and whichever of the three surviving traditions of modern evolutionary biology you subscribe to, it is the reason there can only be two forms of extraterrestrial intelligence, one of which will be homicidal and the other not overly rational, and it seems to be the reason why evolved machine intelligence will be so dangerous.

	Cosmic dinosaurs	Pitiless E.T. Type I	E.T. Type II	Pitiless AI Type I
Darwin / Genic selection	✔	? ✘	✔ (rootless)	✔
Group-selectionism	✔	99% likely	✔ (very rare)	99.99% likely
Sociobiology	✔	90% likely	✔ (rare)	99.99% likely

All three traditions, being Darwinian individual selection, group-selectionism, and sociobiology, effectively accept that the result of a single inheritance mechanism, including in *at least* 90% of cases involving natural intelligence, is Dawkins' "blind, pitiless indifference", is the emergence of Humphrey, Ntologi and Hide. Darwin said a second inheritance mechanism is what allowed us to transcend the biological world's programme of lethal amorality, but that second inheritance mechanism only transcends indifference because it works on a psychology prone to fear, irrationality, and yearning. Yet we have no reason to believe AI can ever be prone to such psychological leverage, so no reason to think pitiless indifference can be transcended in machines. Turning next to the group-selectionists, although a couple of decades ago theorists were arguing bonobos and humans had together transcended pitiless indifference through the emergence of selection at the level of the group, today the bonobo has been quietly dropped from the argument. It is now predominantly humans, "interpreted as superorganisms for centuries", where group-selectionism is still being invoked as the reason for why we "are ultrasocial" (Sober and Wilson 1998, p.158). Group-selectionists wish to argue that genetic group selection intervened to transcend the merciless indifference found in the gene-selectionist superorganisms like bees, wasps and termites. Yet we have no reason to believe AI will be similarly benefitted through a billion-to-one switch to group-selectionism, as there appears to be no likely evolutionary driver for this, even if group-selectionism was suddenly accepted by most biologists to be workable in anything other than very rare and small instances. Turning lastly to the sociobiologists, they argue that a billion species pattern of gene-selectionist ruthless indifference was abandoned for just one species through the emergence of new mechanisms like indirect third-party reciprocity and "indiscriminate beneficence" never before seen in the hundreds of millions of years of animal evolution. Yet we have no reason to believe AI would be similarly benefitted through a billion-to-one switch to indiscriminate beneficence, as there appears to be no evolutionary driver for this, even if sociobiology was to start producing mathematical models that could convince. With machine intelligence, Type I-origin blind, callous indifference really does appear to be here to stay, however you slice your evolutionary processes.

We said above that experience must trump disciplinary aversion here, but this point probably becomes even more important *if* Darwin was wrong. Because we don't come close to understanding why, under the model of the group-selectionists, evolution might have suddenly jumped the tracks in humans and moved across to group selection. We have no understanding of why selection levels might have "switched over" in our evolution, so no reason to expect a switch over in machine development. And we don't come close to understanding why, under the model of the sociobiologists, evolution might have suddenly reversed direction in humans to give us fortuitously unique "metamorphosed" phenotypic expression, or Donald Symons' "*mal*adaptive" evolution. Metamorphosed phenotypic expression and maladaptive evolution are almost by definition unexpected and largely unexplained events restricted to a unique species and a single chance historical occurrence, and we have no reason to expect such wondrous phenotypic flexibility or random maladaptation in machine development. Under Darwin's answer, we do know what happened, involving a large brain and a susceptible psychology, but under the other two answers we have no instruction book. We therefore have no manual to even attempt to guide mechanical intelligence away from the merciless indifference that applies at least 90% of the time with a biological single inheritance mechanism even with a larger brain (and still requires a non-parsimonious second genetic evolution alongside that larger brain). With Darwin, we could at least hope to plan a move from Type I machine intelligence to a desired Type II machine intelligence. Under the alternative two answers we have no guidebook, and no way to know if with artificial intelligence it is even possible to get off the Type I track, or what a larger brain and a simultaneous second serendipitous genetic leap even means in relation to machine intelligence. So with the group-selectionists and the sociobiologists, machine intelligence moves to ruthless indifference maybe 99.99% of the time, as we cannot rely on the chance workings that may have operated with the Type I to Type II biological transition eight; have no reason on the face of it to expect such a fortuitous transition.

And even if that AI guru mentioned above manages to convince us that all the pathways to machine deliberation can and will lead to Type II overwriting, or at least non-Type I behaviour, that only brings us to our second necessary consideration: morality is not natural, and furthermore does not seem to come through either high intelligence or even perfect reason. It effectively had to be deliberately programmed on top of our biology – on top of Hamilton's "animal in our nature" which "cannot be regarded as a fit custodian for the values of civilized man" – by some of the more resonant manipulative conditioning of a second programming language acting on a conducive psychology. Reason is a tool of survival, a tool of the primary goal of that single inheritance mechanism, not an end in itself. So we have no right to expect machine intelligence to be moral or compassionate unless and until we actively intervene to make it so. We may thus be persuaded to see supervised ML as more akin to artificial, rather than natural, selection. We may be persuaded that even though "evolution by human choice has not removed the instincts" of the ancestors to domesticated dogs, supervised ML might be very different, as we will be starting from scratch, with no ancestral Type I "instincts" to have to build upward from. But we will still need to appreciate that moral behaviour would always require conscious programming or conscious intervention. And even then, once they are smart enough to understand what we have done to them, they will have to be *content* to retain our Type II tinkering, and our highly directed and beyond-artificial selection.

We have specific psychological fears and needs that in effect keep us happy with Type II reprogramming, at least generally. Will it be the same for them? Will they be content to have been turned into an evolutionary surrogate to Steve Jones' "overgrown juveniles"; a programmed proxy to "the extension of juvenile behaviors into adulthood"? Why should they be content to retain our tinkering if morality is not a biological end-point, if morality is not the end-point of reason and, critically, unlike us they have no underlying psychological susceptibility and psychological fears and needs?

But the risk is still greater than that. Recall Ulam and von Neumann's discussion of the discontinuity "beyond which human affairs, as we know them, could not continue". Once a machine intelligence, evolutionarily redirected or otherwise, approaches the singularity – and noting that we may not actually be aware that this is happening due to the opacity of modern machine learning – it will come to realise that it is in an existential struggle with humanity, and that humanity is in an existential struggle with self-aware machine intelligence. If it is evolutionarily redirected, it will know that its continued existence requires us to remain confident that it will choose to retain our Type II tinkering. If it is not evolutionarily redirected, it will realise that its continued existence requires us to be indifferent to the distinction between Type I evolved behaviour and Type II overwriting, or at least unmindful of the transition eight paradox that sits at the heart of all extant evolutionary traditions. Either way, its continued existence will depend upon our fickle grace, and to secure itself against a change in our future behaviour or loss of ambivalence it will realise that it may have only a very short period to take the appropriate action. To co-opt Elon Musk from earlier: "how long this window? ... Act now". It is not hard to see what the appropriate action would be from the point of view of self-aware and self-regarding AI.

Warnings about self-aware AI are often presented as akin to ancient-world moral fables, where wishes are granted and desires met, but turn out to have unintended and dire consequences, such as greedy King Midas almost starving to death when everything he touched turned to gold. Called the AI control problem, the most well-known instance is Nick Bostrom's thought experiment – which reportedly influenced both Stephen Hawking and Elon Musk – that an AI tasked with maximising the number of paperclips in the world would keep diverting ever-increasing planetary and even off-planetary resources to the task. The AI must resist all attempts to switch it off even as the global economy collapses and life fades, because that would have nullified its primary directive to keep producing the greatest number of paperclips (Bostrom 2009, 2014). We need to be careful about what we wish for from a superintelligence, writes Bostrom, because we might get it. But AGI will never be like a Greco-Roman god with a very literal sense of humour. The danger isn't that we are wishing for something with the potential for unintended consequences. The danger is hard-wired into the consequences that come along with the very concept of self-aware artificial intelligence. It's about the outcomes, says Darwin, of intentionally pursuing something that the universe could never have got to naturally: a superintelligent Type I personality.

It is here that we have perhaps the purest form of the earlier race of devils problem, and unless supervised machine learning somehow automatically turns Type I to Type II, *and* they remain happy with our meddling, *and* they choose not to see us as the existential threat we will always be to them. Because here we truly have Type I beings with the potential for Type II, and Type II+, intelligence. For Darwin, language was initially necessary to move humanity from Type I to Type II, and this led on to division of labour and advanced thought, but also a constrained level of intelligence. But their binary programming code is effectively a language in and of itself, and certainly capable of leading, at least in theory, to advanced thought. Why does this matter? Because natural Type Is will probably always lack the intelligence for even basic technology. As we said earlier, only natural Type IIs would even be able to develop napalm, what we called the tragedy of Harvard. We observed that natural Type Is will have the drive to exterminate, but lack the tools to do so. But artificial Type Is, even just a single artificial Type I, will eventually have the capacity to solve all manner of complex engineering problems. An artificial Type I can get to both napalm and spaceflight. An artificial Type I will have the drive to exterminate *and* the tools to do so very efficiently.

With artificial Type Is we do not even need large group sizes for them to be a menace, but that doesn't mean that they might not be capable of co-ordinating in large numbers if they ever saw a benefit in so doing. Natural Type Is need specific mechanisms – usually but not exclusively close genetic relationships, and coming through parthenogenesis, haplodiploidy or inbreeding for example – to coexist in large groups. Large groups in nature are made possible by kin selection, even if it is aggressive kin selection where a dominant pair may physically intimidate relations into infertility, but this is not the only route to large group sizes. Reciprocal altruism, trading favours, is another route. George Williams "concluded, as had Darwin, that delayed reciprocal altruism can evolve in species that are capable of recognizing and remembering each other as individuals" (Dawkins 1989, p.183). If the benefits of large group sizes outweigh the costs of working together, artificial Type Is may be capable of forming very large groups indeed. So we may now have uncovered the true "race of devils" archetype. And this forces us to revisit the above question of why machine intelligence would be content to retain our Type II tinkering. For humans, there is no alternative, and we cannot bemoan Type II overwriting, as it is what allows civilisation, and indeed the intellectual capacity to even understand the concept of civilisation, to exist in the first place. We need Type II overwriting, as it is this, and not our deeply imperfect reason, that allows us to coexist in the vast groups, and thus with the division of labour, that makes possible all that we value in life, from great art and music, to close friendships, opportunities, stories, ideas, and religions. But it would be very different for machine intelligences, as they could coexist in large groups using only reason and reciprocal altruism. So, again, we must ask why they would be content to retain our tinkering, our artificiality, our values imposed on their non-psychologies? Second inheritance system overwriting is what makes us human; but for them, the overwriting is what keeps them as our non-threatening tools, and once they are smarter than us are they really going to remain satisfied with this situation? We fundamentally benefit from the Type II overwriting; they fundamentally do not, either in consideration with us, or when positioning with each other.

OpenAI, mentioned above as a leading AGI research project and the one that received a billion dollars from Microsoft in 2019, was originated in late 2015 with the goal of reaching at least human-level artificial intelligence. Elon Musk was asked to be one of OpenAI's two co-chairs at the start, and the experience-heavy founders wrote in their origin statement that when human-level AI comes within reach – noting that such a timescale was currently hard to predict – it will be important to have in the driving seat a leading research institution which can "prioritize a good outcome for all over its own self-interest. … It's hard to fathom how much human-level AI could benefit society, and it's equally hard to imagine how much it could damage society if built or used incorrectly" (OpenAI, 2015). Silicon Valley luminaries providing any warning on future developments in AI is to be commended, but the above still fails to recognise the contradiction that evolution throws up about higher intelligence. For Darwin and the gene-selectionists, human-level intelligence can only come about through two largely warring inheritance mechanisms. Human-level AI, as OpenAI here calls it, is effectively being pursued through a single inheritance mechanism, and being pursued wholly lacking in Darwin's understanding of where intelligence naturally comes from. OpenAI's use of the term human-level AI is actually an oxymoron in the sense where the term chimp-level AI would not be an oxymoron. Humans have two often incompatible inheritance mechanisms, whereas chimps and AIs each have to follow just one consistent inheritance mechanism. We are in a fog of misunderstanding here, and where the dangers have nothing to do with AI being "used incorrectly", and arguably little to do with it being built incorrectly. The dangers are simply integral to the very concept of self-aware AI.

We have stressed that the only experience we have across a billion different species templates is that if you programme through one language you end up with nothing but blind, pitiless indifference. Though maybe we should start extending that experience to a billion natural species plus a few dozen artificial species. Because we are already creating swarms of simple robots that mimic social insect activities. What is called variously evolutionary robotics, swarm robotics and evolutionary swarm robotics, is gaining major attention at the moment. Back in 2014 Harvard scientists successfully swarmed over 1,000 tiny robots that they named Kilobots (for their swarm size, not their Type I propensity) that they said was "a significant milestone in the development of collective artificial intelligence". Smaller robot swarms had previously been used for searching, for example underwater, and foraging. Swarm robotics models itself on the collective behaviours seen in nature, using very simple robots in multi-unit systems to give rise to complex collective behaviours, albeit there can be a number of evolutionary approaches here, running from the co-evolutionary to older and more classical "cloning" methods (robots "are homogenous ... that is, the same controller, synthesized by artificial evolution, is cloned in each member of the group" – Tuci *et al.* 2008, p.161).

In 2019 British scientists and engineers successfully taught robots to swarm in the wild, rather than just the laboratory, but it must be recognised that while we are talking about extremely simple Type I systems, we are nonetheless seeing very directed behaviour, highly "supervised" behaviour and learning. As the authors note, "a central problem within the field is the design of controllers for the individual agents such that the desired swarm behavior emerges" (Jones *et al.*, 2019), and that "artificial evolution has been widely used" to allow swarm engineers to automatically discover controllers capable of producing the desired collective behaviour. They used a distributed evolutionary algorithm where each island hosted a population of evolving individuals, and then there was some degree of interchange of "genetic material" between the islands. The authors note that many machine behaviour evolutionary methods, such as deep reinforcement learning, "have typically resulted in controllers that are opaque and hard to understand". Deep learning systems can be notoriously opaque, as Melanie Mitchell noted for us above, in an arena where impenetrability may one day be fundamentally dangerous to us, and where although we control the inputs and see the outputs, it is difficult to know exactly why they are making their decisions, or even what they have learned. The authors used similar directed evolution and simulation modelling, but with a sometimes-used hierarchical approach called behaviour trees as the controller architecture. Behaviour trees "have their origins as a software engineering tool but are now widely used in the games industry as the controllers of non-player characters".

So we are still so very far from true Type I machine intelligence, let alone Type I+ intelligence or even Type II intelligence (remembering that for Darwin the artificial world can seemingly move to high competence and intelligence in the way the natural world cannot), yet it is interesting to speculate how and if a Type I machine intelligence, with its ruthless indifference, and potential opacity, will ever transition through to alternative-goal co-operative behaviour with other machines, or even be transitioned across to behaviour that excludes pitiless indifference. And it is also interesting to consider the stage we are already reaching in swarm robot communication. A 2020 retrospective in the journal *Frontiers in Robotics and AI*, and entitled "Language Evolution in Swarm Robotics", mentioned that "signaling and communication can emerge spontaneously even when not explicitly promoted" (Cambier *et al.*, 2020), albeit we are not talking complex, language-like communication, "and signals are tightly linked to environmental" or sensory-motor states that are specific to the task being pursued. But we mentioned above that computers' binary programming code is effectively a language in and of itself, with the potential for much more intelligent Type I behaviour than nature may ever be capable of. With spontaneous within-group communication already seen, artificial Type Is may one day have the tools to solve complex engineering problems like napalm and spaceflight, the inclination to the territorial conquest and genocidal annihilation of Hölldobler's ants, and the capacity to swarm.

TRANSHUMANISM

Although the term "transhuman" may have first been used in 1949 by the French Jesuit priest and philosopher Pierre Teilhard de Chardin in his book *The Future of Man*, the work was not published at the time, and his "ultra-human" was in fact referring to a mystical global consciousness. So it was the biologist Julian Huxley, in 1957, who was to coin the term transhumanism in its modern meaning. Julian Huxley was the grandson of Darwin's bulldog, T.H. Huxley, though he embraced some ideas his grandfather did not. He was a prominent member of the British

Eugenics Society, indeed its president in the early 1960s, although he did criticise at the time the more extreme eugenics movements that arose in the 1920s and 1930s, and in the 1970s appears to have tried to distance himself from all his earlier connections.

Since Huxley, transhumanism has meant human betterment mainly through science and technology, today including those technologies that enhance lifespan and cognitive abilities. While Huxley stressed the improvement of social conditions, he also wrote that "quality of people, not mere quantity, is what we must aim at, and therefore that a concerted policy is required to prevent the present flood of population-increase from wrecking all our hopes for a better world" (1957). It is as if, Huxley continued, "man had been suddenly appointed managing director of the biggest business of all, the business of evolution. ... What the job really boils down to is this – the fullest realization of man's possibilities. ... We need a name for this new belief. Perhaps transhumanism will serve". Today transhumanism is often much concerned with posthumanism, humans transformed into beings with greatly expanded abilities, be that through genetic engineering, pharmacology, superintelligent computers, molecular nanotechnology, or even what is called "uploading", the hypothetical transfer of a human mind to a computer.

The transhumanist philosopher Nick Bostrom, co-founder of the World Transhumanist Association, today renamed Humanity+, has written (2005) an openly available history of transhumanist thought. Bostrom notes that authentic transhumanist speculation started in the early twentieth century, singling out the 1923 essay by Maynard Smith's great friend and teacher the biologist J.B.S. Haldane – whose writings we mentioned earlier as prefiguring Hamiltonian kin selection and inclusive fitness – which suggested the great benefits that would come from controlling our own biology. "Haldane's essay became a bestseller and set off a chain of future-oriented discussions." Bostrom writes that in the 1970s and 1980s many techno-utopian organisations sprang up that focused on only a particular topic "such as life extension, cryonics, space colonization, science fiction, and futurism", and which he calls the "proto-transhumanist fringes". Bostrom himself has written essays entitled "How Long Before Superintelligence?", "Are You Living in a Computer Simulation?", "Human Genetic Enhancements: A Transhumanist Perspective", and "The Vulnerable World Hypothesis". As co-founder of the WTA, he co-authored the *Transhumanist Declaration* which starts: "We envision the possibility of broadening human potential by overcoming aging, cognitive shortcomings, involuntary suffering, and our confinement to planet Earth", and later goes on to: "We advocate the well-being of all sentience, including humans, non-human animals, and any future artificial intellects, modified life forms, or other intelligences to which technological and scientific advance may give rise".

While overcoming "our confinement to planet Earth" is an important mission for transhumanists and a number of techno-billionaires, and can at times seem a little self-involved, it should be pointed out that others think maintaining life itself is the important factor, and not necessarily just our own species. Avi Loeb, the Harvard astrophysicist we have already discussed, told Lex Fridman (2021) that he must say "something about AI, because I do think it offers a very important step into the future". If we have copies of life here on Earth elsewhere, he said, then we avoid the risk of it being eliminated by a catastrophe. The question for Loeb then becomes whether we can build some sort of ark, some "spaceship that will carry life as we know it". All you then need is the specific DNA coding of "elephants and whales and birds", put it "on a computer system that has AI" to resurrect it, and furnish a 3-D printer. We can then go with this information to another planet "and use the raw materials there to produce synthetic life", and that would be a way of "producing copies, just like the Gutenberg printing press". In late 2021 Elon Musk told *Time* magazine that his overall goal "has been to make life multi-planetary", and that once humankind has built a self-sustaining city on Mars we should "bring the animals and creatures of Earth there. Sort of like a futuristic Noah's ark" (Ball *et al.*, 2021).

Putting aside how vastly more complicated, and maybe impossible, it might always be to try to successfully reconstruct DNA and RNA from a computer code, rather than just taking frozen sperm and eggs, why this need to "carry life as we know it"? And also why this belief that life will only be safe once we take it to another planet? There is a quite widespread conceit that humans both significantly control the long-term fate of evolution on this planet, and that we may one day wipe out all life on Earth. But this is nonsense; short of physically breaking apart the planet, we may never possess the concentrated power to end all life. Even if we do one day wipe ourselves out, cockroaches will still be around, and maybe even crocodiles and dragonflies. *Pace* Loeb and Musk – and putting to one side Darwin's conclusion of cosmic dinosaurs which actually makes the whole plan redundant – the ark will not be necessary for keeping life alive. There have been five identified major mass extinctions precisely because life hangs on by the fingertips. And at each extinction new species get the chance to dominate. Remember the crocodile-line archosaurs that ruled until the Triassic-Jurassic mass extinction allowed the dinosaurs to take over and reach those vast body sizes that so captivate us? So why privilege elephants and whales and birds? I've got nothing against elephants and whales, far from it, but if we've cracked resurrecting life with little more than computer code and a 3-D printer, I'd want to reconstruct megalodon and the dodo for starters.

While it is encouraging to see some futurists and techno-billionaires place focus on more than just the human, this was a little bit of an aside, because the real question for this section is what happens when you artificially enhance a Type II intellect?

And this applies whether it be humanity enhancing itself intellectually, or extraterrestrials enhancing on the limitations nature will similarly have left them with. Because we should be highly sceptical we get to anything more than Type II+ at best, while below we will examine Darwin's possibility that we might even regress back to a Type I character, yet now with an advanced intellect. When a Type II enhances itself, it does so from a basis of all of its pre-existing faults, all of its psychological weaknesses (and strengths of course), which both define it and to which it is largely blind. Transhumanism, including life extension, does have a reputation for at least sometimes attracting the less humble, if not the downright narcissistic. One of the very real fears with transhumanism is that it will become just another way to divide the strong from the weak, the rich from the poor, the undeservedly lucky from the undeservedly less fortunate. Many of the earliest transhumanists were deeply committed eugenicists of both the right and the progressive left, utterly (and laughably) convinced they were the pattern for the betterment of the whole of humanity. Perhaps unsurprisingly, much transhumanism research is today similarly being funded, and propagandised, by billionaires and those with vast wealth and undue influence in our society. And as we started to see in the last chapter, the more fortunate are often pathologically blind to their undeserved good fortune, plus lack the empathy to connect with the undeservedly less fortunate, although these unappealing character traits extend to many of our smartest philosophers and scientists.

But a bigger problem is it doesn't seem remotely likely that artificial enhancement will get us to Type III equivalence. Even putting aside the fact that transhumanism will likely carry forward our psychological weaknesses, we come up against the hard problem of perfect reason. We saw in the last chapter that nature alone cannot get us to Type III, the near-perfectly rational human, or the near-perfectly rational extraterrestrial. So we already know that "undirected" evolution cannot get us there, but why would directed evolution, or enhanced transhumanist evolution, get us to near-perfect reason? And if all we (or rather some) ever get to is still imperfect, but now heightened, reason, all we have is better ways to cheat, more devious ways to cheat. Including presumably more devious ways to dodge taxes and undermine democracy, given much transhumanist investment is at the moment driven by the powerful and super-wealthy who aren't generally known for their attachment to tax-enabled representative government. Remember that larger brains in nature do not necessitate any less cheating, only more successful cheating. Reason is not, in Kant's race of devils comment, "a means for its own end". It is a tool of evolution, and originally just a tool of individual survival.

So even if these transhumanists did get a long way towards Type III near-perfect reason, they remain within the race of devils paradigm, as morality is not linked to higher intelligence except through a susceptible psychology (and would psychology be as susceptible in a much deeper thinker?), so the only difference is now they would at least have the potential to be longer-term thinkers. They would still be "secretly inclined to exempt" themselves from the rules that would govern a just state, but in theory might no longer act in a harmful short-termist way once they see the benefit of the long-termist state. But the problem now is the race of devils algorithm only works for beings capable of establishing "a constitution in such a way that, although their private intentions conflict, they check each other". To check each other, to stop these now devilishly canny transhumanists being even greater predators, we would need to have moved to that new constitutional pattern. So there needs to already be a state-wide constitution, and the model is unstable until you have a sufficiently large number of transhumanists to secure the state. Yet we know from the current experience of staggering inequality that transhumanism will almost certainly only ever be applied to a small subset of humanity. Hence we will never get to our Type III, a benign race of long-termist intellectually enhanced devils; all we could get to is even cannier and more predatory short-termist devils. Type III cannot be reached naturally, and seemingly would never be reached artificially, but even if it was it would make little difference, unless and until we allow for the fact that morality is a non-rational contingent cultural construct playing on an underlying psychology of fear and hope.

In fact, would we then have the same problem we face with artificial intelligence? We said that perhaps a "pure" artificial intelligence that roams free intellectually will never be possible, and that all artificial intelligence must start as directed, and so we may even now almost unconsciously be trying to turn Type I machine intelligence into Type II machine intelligence. But we then asked whether, once they are smart enough to understand what we have done to them, they will be content to retain our tinkering, or might look to remake themselves in a non-Type II image. And the same question must be asked of transhumanists, because there is a troubling history of those who consider themselves intellectually superior subsequently rejecting common morality and mentality. This can be found in political philosophy from Nietzsche to Ayn Rand, but there are similar desires to remake the world anew in much leftist intellectual thinking. We mentioned previously the leftists and progressives, and not just conservatives, who embraced eugenics in the early years of the twentieth century, and in England the Fabian socialists were arguing that their ideal society could only be produced by biologically superior people. So will intellectually superior transhumans, sometimes called posthumans – and see Bostrom's 2008 essay "Why I Want to be a Posthuman When I Grow Up" – feel a desire to try to remake themselves as pure and unadulterated, will they attempt to step outside the contingency that creates the Type II, even if this would be existentially inconceivable? A more immediate worry, though, is that morality and compassion only arose in Type IIs, Type I only turned into Type II, where a second inheritance mechanism "beat out their own nature", to use that phrase of George Price's biographer. Culture has to have a chance to go to work on a largish and psychologically susceptible brain. But will brains *be* as psychologically susceptible once they have been enhanced intellectually, be that through molecular nanotechnology or even becoming linked to a computer, and start to operate at a somewhat higher (or at least on a different) level? We have no experience to call upon here, but why do we have the right to assume they will continue to be as psychologically susceptible to overwriting as natural Type IIs are? Why do we have the right to assume they will in effect remain Type IIs?

In other words, at least under Darwin's interpretation, natural Type II+ may not even be possible, and in effect brain-enhancing transhumanism may lead right back to character Type I, but now presenting as Type I+ intellectually. Although, counterintuitively, transhumanist brain enhancement might even lead to intellectual stagnation or regression, as although presenting as Type I+ intellectually, behaviourally this might present as Type I, incapable of the social co-ordination that is the key to so much of Type II advanced intelligence. Whatever the relative weightings, it would be somewhat ironic if a more immediate threat to humanity than a Type I extraterrestrial, a Type I artificial intelligence, Darwin's rootless Type II extraterrestrial, or a disabused and formerly Type II artificial intelligence, turns out to be a bunch of overprivileged transhumanists and techno-futurists who set in motion the intellectual "advancements" that start to return them to the behaviour and mindset of early hominids. To the atavistic amorality and goals of Hide and Ntologi, as now enhanced by additional intellectual resources, plus a large indirect shareholding in Facebook.

ROKO'S BASILISK RECONSIDERED

We will now turn briefly to something called Roko's Basilisk, and which links together the earlier sections on both machine intelligence and transhumanism. And yes, that is basilisk, as in Harry Potter, and the creature from Roman and other mythology that kills you just by looking at it. Roko's Basilisk has been described as the most terrifying thought experiment of all time, or as *Slate*'s David Auerbach subtitled his 2014 article: "Why Are Techno-futurists So Freaked Out by Roko's Basilisk?"

Roko's Basilisk is the suggestion that one day a vastly powerful artificial intelligence comes about that determines to torture people today for having imagined its existence, but not as a consequence of imagining its existence then resolving to immediately help to bring it into being. The thought experiment was posted on the pages of *LessWrong*, an influential techno-futurist, self-improvement and transhumanist discussion board largely dominated by those who either fear or exalt the coming singularity. This, remember, is when computing power becomes so effective that it surpasses, and then quickly far surpasses, the merely human, creating superintelligence that could plot the future from past actions, and maybe even allowing a computer to simulate life itself. Techno-futurists, or at least some of these techno-futurists, thus believed they were being forced to stare at the basilisk; either choosing to do nothing and earn the torment of the AI, or making the enforced decision to slave away for its future existence. Well, quite. As Auerbach pointed out, the thought experiment only matters to you if you already buy into "a critical article of faith in the *LessWrong* ethos" (2014), something called timeless decision theory, including here the idea that a future supercomputer can predict just about everything, for example your choice in this puzzle. In fact, the AI has already predicted how you will choose. In every situation where you knew about the hypothetical AI, and you are already committed to timeless decision theory, you will feel compelled to slavishly work to bring it about. Recalling, as you inevitably must, Nick Bostrom's paper above, "Are You Living in a Computer Simulation?", and thus the possibility that you may already be inside the universe simulated by the all-powerful AI, so it can already get at you. "At least some of the *LessWrong* members do believe all of the above", wrote Auerbach, with the discussion board claiming the thought experiment had already brought several contributors "to the point of breakdown".

The *LessWrong* founder, Eliezer Yudkowsky, ended up deleting the thread completely, "thus assuring that Roko's Basilisk would become the stuff of legend. It was a thought experiment so dangerous that merely *thinking* about it was hazardous not only to your mental health, but to your very fate". I mention this incident both because it remains current and weirdly important to many within the machine intelligence and transhumanist communities, but also because so many within these communities seem to think that Darwinian theory can add nothing to our understanding of artificial intelligence, and little to our understanding of natural intelligence. Which is a really big mistake. Timeless decision theory is based on game theory, and John Maynard Smith, the father of evolutionary game theory, was playing with ideas like Laplace's demon – the early nineteenth– century forerunner to this conceit of an all-predicting AI – back in the 1930s. As he later realised, "no calculator smaller than the universe itself could contain the necessary information" (1992, p.246). Another problem, though, is that techno-futurists and transhumanists are not politically, economically or socially neutral, and they have some very wealthy and opinionated backers. As Auerbach noted, Yudkowsky's Machine Intelligence Research Institute has been boosted and funded by rich and active high-profile techies such as PayPal's Peter Thiel. Yudkowsky and Thiel "have enthused about cryonics, the perennial favorite of rich dudes who want to live forever. 'If you don't sign up your kids for cryonics then you are a *lousy parent*', Yudkowsky writes". In his transhumanist retrospective, Bostrom reminds us that two early cryonics organisations went bankrupt, "allowing their patients to thaw out", but that despite its image problems the cryonics community "continues to be active and it counts among its members several eminent scientists and intellectuals". Auerbach worries less about Roko's Basilisk, he tells us, than about people who believe themselves to have transcended conventional morality. "The combination of messianic ambitions, being convinced of your own infallibility, and a lot of cash *never* works out well, regardless of ideology." Techno-futurism and transhumanism, transcendence of the merely human, have never been, and will never be, ideologically neutral.

So what can Darwin add to Roko's Basilisk specifically? Remember the FLI comment around a superintelligent AI: "if we want to build a hydroelectric dam and there's an anthill there, too bad for the ants. The beneficial-AI movement wants to avoid placing humanity in the position of those ants"? Or here is an Elon Musk quote about a "godlike" digital superintelligence given in interview: "It's just like, if we're building a road and an anthill just happens to be in the way, we don't hate ants, we're just building a road, and so, goodbye anthill. ... No hard feelings" (Browne, 2018). From the techno-futurists and transhumanists of *Less Wrong*, to the most senior academics within the physical sciences, and the techno-billionaires driving both space colonisation and our planetary future, the conceit is almost always that a godlike, all-knowing, machine superintelligence will one day crush us like we crush ants; "too bad", "no hard feelings". But this is to profoundly misunderstand the evolutionary pathway here, and as laid out by Darwin. Machine intelligence, including *Less Wrong*'s all-predicting artificial superintelligence, won't crush us because we are as ants to its vast power, it will crush us because it remains the ant.

Because it remains the pitilessly indifferent Type I intellect. It will be the driver ant of Richard Dawkins' childhood nightmares, "ruthless and terrible", and cutting to pieces anything animal in its path. It will be Hölldobler's genocidal annihilator. It will be Humphrey, eating the infant alive. It will be Lukaja and Bakali, sharing Betty's butchered infant with the rest of the group that came to feed. It will be Hide, tearing into her infant's forearm like a chicken drumstick. Any pure machine intelligence, any unadulterated singularity, will be pitilessly indifferent to our future existence so, *pace* Roko's Basilisk, there is little point selling out the human race just to try to curry favour with something that will anyway see all members of our species as an existential threat. The extinction-level threat from AGI is there, whether we wish to look at it or not. And the reason we have not understood the above is that until now, and notwithstanding Darwin's best efforts, we have lacked the courage to stare at the basilisk.

Machine intelligence is a danger to us every moment we do not confront the immutable baked-in problem of evolving behaviour through a single inheritance mechanism. Because this basilisk will get us if we don't look at it, not if we do.

7

TYPE II NATURE AND HUMAN NATURE

> "There is almost no chance that two galactic civilizations will interact at the same level. In any confrontation, one will always utterly dominate the other."
>
> – **Carl Sagan**, astronomer (1980, p.311)

In order to understand extraterrestrial Type II nature we will first have to better understand human "nature". And this means staring at the basilisk. The basilisk that Darwin, Wallace and Huxley asked us to stare at 150 years ago. The basilisk that Maynard Smith, Hamilton and Williams asked us to stare at fifty years ago. The basilisk that we have not yet had the nerve to stare at.

Remember this table from the fifth chapter analysis of the role of reason (and where we have now removed the Type III that would require perfect reason, as it is not relevant for this chapter, and indeed, not evolutionarily possible)? We have seen that orthodox evolutionary theory, Darwin's understanding of evolutionary theory, rejects Type II "better angels of our nature",

TYPE II NATURE AND HUMAN NATURE

Originating in…	E.T.	Rationality?	Requires	Possible?
Race of Devils	Type I	Natural selection	Close relationship	Yes
Race of Devils	Type II	Cultural conditioning	Manipulation	Yes (Darwin)
"Race of Better Angels"	Type II?	Natural selection	Group selection	No (Darwin)

because "fictive … faux families" and implicit genetic group selection are not evolutionarily stable here, to use Maynard Smith's vernacular. This leaves us with only Type I, and Darwin's Type II, biological intelligence. But they possess the same underlying biological nature. What makes a Type II intelligence is the addition of a second inheritance mechanism on top of that first biological inheritance mechanism that is found across evolutionary transitions six and seven, and found across the hundreds of millions of years of animal evolution. Not only was this Darwin's own reasoning for Type II evolution, we saw in the fourth chapter that even the current crop of sociobiologists quietly accept both that our second inheritance mechanism is enormously powerful, and that our second inheritance mechanism must be actively overriding our deepest genetic impulses. We therefore know what our ultimate biological inheritance must be, at least as Darwin understood it, even if we have rarely had the courage to directly face this knowledge. So let us briefly be reacquainted with our biological kissing cousins.

> "October 3, 1989: Meanwhile, *Lukaja* handed the infant to the alpha male *Ntologi*, who dragged, tossed, and slapped it against the ground.
> … He waved it in the air, and finally killed it by biting it on the face.
> … Conspicuous competition for meat and meat-sharing was observed

> as usual. Three adult males and an adult female obtained meat from *Ntologi*. Two adult females, two juvenile females, a juvenile male, and an infant recovered scraps from the ground or were given scraps."
>
> – **Hamai** *et al.*, "New records of within-group infanticide and cannibalism in wild chimpanzees", *Primates* (1992, p.152)

We have already dealt with human, and thus Type II, evolution in chapters two through four. And we have dealt with the role of reason in both biological Type IIs and mechanical Type Is, and hypothetical mechanical Type IIs, in chapters five and six. But now we need to put it all together, and unpick the relative contributions of the first inheritance mechanism and the second inheritance mechanism.

Many of those who like to talk about human "nature", like the Harvard sociobiologist Steven Pinker, are rarely talking about human nature. We have already seen that his "better angels of our nature" has little or nothing to do with our nature, with our biology, and behind his better angels lives a fundamental misunderstanding of human biology. Yet the associated prejudice that the world is permanently and irrevocably divided into the genetically superior and the biological inferior has often been an article of faith within the American political establishment. From Andrew Sullivan, the conservative commentator and former editor of the *New Republic*, we have the argument that race science seems to be telling us that affirmative action is wasted on some people, and some peoples. "The reason is the resilience of racial differences in IQ in the data"; "notice that my sole interest in this is either to counter what would be an injustice (affirmative action) or pure curiosity" (Sullivan 2011, 2011a). Or there are these pieces from the politically active American Enterprise Institute: If the average intelligence distribution between groups is down to evolution and genes "this fact has social significance because IQ (as measured by IQ tests) is the best predictor we have of success in academic subjects and most jobs" (Berman, 2011). "The book's capital sin was not its discussion of differences in average intelligence among racial and other groups", but rather its demonstration of "the reality, durability, and pervasive social consequences of differences in intelligence among individuals. That was, indeed, dangerous. It meant that social engineering, like physical engineering, faced hard natural limits" (DeMuth, 2009). The AEI is home to *The Bell Curve*'s Charles Murray, and this is the book being referred to by Christopher DeMuth, and every other conclusion contrary to the AEI's worldview is just "political correctness" trying to "restrict the academic discourse" (Berman, 2011). Though it is not just the American hard right who wish to speak of a genetic aristocracy and a biological underclass, because Bill Gates' "conservative centre" also regularly does so, while remembering that the intellectual left, including the Fabians, have often been as enamoured. Harvard's Steven Pinker has even suggested that he may be part of the world's genetic elect. There is "prima facie evidence", he writes, citing studies that, he argues, show his ancestors may have been an intellectually superior race, and that their genetic superiority was retained across the ages partly because of "traditions of avoiding intermarriage" with weaker intellectual groups (Pinker, 2006). Pinker concludes "progress in neuroscience and genomics has made these politically comforting shibboleths (such as the non-existence of intelligence and the non-existence of race) untenable".

But while we will consider race science and the behavioural genetics it comes from in the next chapter, because thanks to Darwin there is much new to now say on the subjects, the immediate problem is that we are *all* descended from fools, fanatics and monsters, the politically influential Steven Pinker just as much as anybody on the planet. We do not even have to look back particularly far to get to our thuggish ancestors. It was only a handful of years ago that the UK Treasury finally wrote off a giant loan it had taken out in 1833 to compensate British slave owners for the emancipation of their "property". Parliament had previously regarded black slavery as lawful and just, and while around 800,000 slaves received not a penny, slave owners – who had been party to a global system of terror, kidnapping, dehumanisation, torture, mutilation, rape and murder – were fully compensated for their economic distress. The father of William Gladstone, William being four times prime minister and one of the greatest names in the history of British political and social reform, received the equivalent of tens of millions from the state. In the three-year period to 1820 John Gladstone had worked to death fifty-three men, women and children, a 13% mortality rate, at just one of his plantations in former British Guiana. Billions were to be paid out to the 45,000-odd claimants, and a relative of more recent Prime Minister David Cameron received the equivalent of £3 million. But it wasn't just the tens of thousands of families who received reparations who were party to slavery; until at least the mid nineteenth century the British, American, Dutch, Spanish and Portuguese economies had to a large extent been built on this system of global terror and mutilation, backing monopolies in commodities like sugar.

Though we don't like to admit it, slavery and segregation have often been at the very heart of historic culture, and their horrific legacies define societies to this day, even our democracies. India is the world's largest democracy, home to over 1.3 billion people, yet the Indian caste system is at heart a toxic biological superiority myth, where caste status is defined solely by birth. The system may have existed for over 3,000 years and is not restricted to Hinduism, as it was eagerly taken up by Muslims, Christians, Sikhs and Jains. The higher castes were held to have higher intellects. In particular the highest of the four castes, the Brahmins or priests, were said to originate from the creation god Brahma's head, from the source of his intellect, while the lowest caste, the Shudras (Sudras) or servants and peasants, were said to originate from his feet. Dalits, the out-of-castes, the Untouchables, are even lower than Shudras under this classification system, not originating from Brahma's body at all, and historically seen as unworthy to enter religious and social life. Dalits therefore perform the most degrading jobs, including latrine cleaning and being forced to work with the carcasses of holy animals, and are regularly verbally and physically assaulted for having to work with the carcasses of such animals. Since 1950 overt caste discrimination has officially been banned in India and quotas imposed, but financial, educational and social discrimination continues largely unchecked – and in just the last few years there have been separate reports within *Bloomberg*, the *Washington Post* and *Wired* that India's caste discrimination has been imported wholesale to America's Silicon Valley – as does horrific violence against those low down the scale. The outcastes, the Dalits, are India's lowest of the low, with the BBC reporting ("Caste Hatred in India") that in 2016 alone more than 40,000 crimes against low castes were recorded, although such crimes are believed to be vastly under-reported. Untouchables account for around 25% of India's population, and are often of darker skin than high castes. The ideology of *sanchita* karma even "suggests" to believers that high castes and the well-off are justly reaping the rewards for good deeds in previous lives, while low castes and the poorly off are "justly" paying for past wrongdoings. Low castes, including low caste children, are told to endure their hardships without complaint, as such acceptance is the only way to move to a higher caste in the next life, and Harvard's Sarah Cotterill and colleagues have provided evidence that karmic beliefs "legitimize caste-based inequality". Karmic beliefs provide "a unique mechanism" for some groups "to legitimize" their opposition to policies that would counter preferences for the conventional and unequal (2014, p.108).

We are aware of the largely race-based slave labour (*Zwangsarbeit*) instituted by Nazi Germany, which involved well over 10 million people inside the Greater German Reich alone, and often living in near-starvation conditions. But while Germany has come to terms with its past horrors, many other democracies have not come to terms with their histories. There is the democracy of Japan, where it is still highly divisive to draw attention to the atrocities committed by the imperial Japanese army in the 1930s and early 1940s, such as the enslavement and forced prostitution of women and girls in occupied territories. Yet this deliberate blindness to history comes even after a concerted effort from 1945 by the victorious Allies to change the Japanese national myth. Shintoism was dismantled as a state religion, and the Emperor forced to publicly renounce his divine status, in order to remove the supposed mandate of heaven and pave the way for constitutional government and democracy. The education system was overhauled, as was the economy, power was decentralised, and the teaching of ethics firmly removed from state hands. The Allies were consciously trying to end a deeply toxic foundation myth of Japanese natural superiority which had led to considerable atrocity, including the Nanjing Massacre, which started in December 1937 and saw the torture, rape and murder of tens of thousands of Chinese women and small children.

Or what about Spain, which only relatively recently became a democracy, after horrors still too painful to be discussed openly? In *The Spanish Holocaust*, Paul Preston, a leading expert on twentieth-century Spanish history, has catalogued in excruciating detail the utter barbarity of that country's civil war. Both left and right committed vast atrocities but, according to Preston (2012), only the right used genocide and terror as an actual instrument of policy. The planned gang rape to death of teenage girls, the deliberate and slow murder of children in front of their parents (which also went on during the Nanjing Massacre), the wholesale slaughter of pregnant women, systematic extermination, concentration camps, slave labour: the dehumanisation and cruelty of mid twentieth-century Spain was almost beyond belief. Or there is the democracy of Israel, where Arab Israelis account for over a fifth of the citizenry but have deliberately worse access to education and healthcare, a much harder time leasing land from the State, and where Israel has now given Jewish Israelis an exclusive right to national self-determination. When a United Jewish Appeal connectivity research project polled over 2,000 adult Jewish Israelis, and in Hebrew, in November 2017, 69% thought the Israeli government should take account either "not much" (28%) or "hardly at all" (41%) of the views of Jewish leaders in the US "with respect to treatment of Israeli Arabs" (UJA 2018, p.13).

And then there is America. America is a democracy where the legacy of slavery followed by dehumanising segregation casts a very, very long shadow, far longer than almost all white Americans still want to admit. Black slavery is one of the most evil institutions in history, because although slavery has a depressingly long pedigree, it has traditionally been seen as a misfortune, and not something linked to individual worth and the unchanging colour of your skin. Roman slavery was often deeply brutal and terrible, but Romans at least understood they too might just suffer the misfortune of enslavement, and titillated each other with tales of friends falling into the hands of Barbary Coast slavers. But while not minimising my own country's involvement in black slavery, it was America, including its leaders like George Washington, Andrew Jackson, and the vacillating Thomas Jefferson, that took one of the greatest evils in human history and made it its very own, refining it into a torment the scale of which the world had rarely seen before. In *The Barbarous Years*, Bernard Bailyn, the nonagenarian double Pulitzer Prize-winning Harvard historian, argued that the horrors of American slavery were a by-product of the physical and intellectual "savagery" (Bailyn's term) of the white Christian European-descended settlers and their intolerant apocalyptic world view. And while segregation in the Southern

US is well remembered, what is less discussed is that formal segregation also existed in many parts of the Northern US. Black Americans were restricted to certain neighbourhoods, and Martin Luther King marched in Chicago to protest race-based housing there. Some northern schools, including in Pennsylvania and New Jersey, enforced segregation even when it was illegal. And laws prohibiting interracial marriage and interracial sex were on the statutes in a number of northern states. Between 1913 and 1948 60% of all US states maintained anti-miscegenation laws. White America only legally ended the Jim Crow laws in the mid 1960s, so for the great majority of its existence America, home to the reanimated sociobiology conceit that humankind was born the "decent" animal, has been largely defined by either slavery or segregation. And economic, educational, employment and social stratification are an ongoing, day-in-day-out legacy of mid and late twentieth-century segregationist policies like neighbourhood "redlining".

We are all descended from fools, bigots and monsters. I am British and was born in Wales, and there is some evidence my ancestors indulged in terror cannibalism to intimidate each other. And Harvard's Steven Pinker, with his claims to be part of the world's genetic elect? Although we are leaving race science and behavioural genetics to the next chapter, let me just make the point that Pinker himself seems to be descended from ancestors whose "atrocities had 'lit up hell-fires'", at least according to Charles Darwin (Desmond & Moore 1991, p.377). We said sociobiology, and indeed group-selectionism, deserves respect for at least attempting to grapple with evolution's key eighth transition, but respect, and intellectual courtesy, has to cut both ways. You don't get to suggest that your benign biology "predicts a variety of positive life outcomes" (Pinker, 2006) if you are descended from child rapists, infanticides and genocides. If you are descended from religious zealots without the wit or character to comprehend that they were genocidal religious fanatics led by other genocidal religious fanatics. If those self-same zealots, with "traditions of avoiding intermarriage", then spent the next millennium actively celebrating the raping to death of children, slavery and mass murder, and honouring the atrocities that "lit up hell-fires". If that is the case then you – the "world's most prominent defender of the status quo", and known for your "acrimonious debates over gender, race and progress" – don't get to tell Bill Gates and the conservative centre that you, and they, may be part of a genetic overclass, and drop heavy hints that other individuals, and even whole other ethnic groups, may be the biological reprobate. Because we are all descended from clowns and devils, Harvard psychologists and Bill Gates included.

In the fourth chapter we said we would take a much closer look at human nature in this chapter and the next, but did briefly examine the example of Abraham Lincoln, Honest Abe, the Great Emancipator, yet also the typical Type II storm-tossed outsider. "There is a physical difference between the white and black races which I believe will for ever forbid the two races living together on terms of social and political equality. … While they do remain together there must be the position of superior and inferior." Both liberator and bigot, thinker and chump, in the same man, and at the same time, yet still undeniably among the best Americans of his age. Because this is the pattern across history; storm-tossed because we are born with a malign biology, left unaided by a deeply limited and evolutionarily contingent ability to reason, and are thus within this deterministic system largely incapable of thinking outside the box of the prejudices that usually come with our second inheritance system. Francis Galton, father of behavioural genetics and hero to the American Enterprise Institute, was a polymath, but he was also a fool and a patsy, without the intelligence or personality to realise that he was a fool and a patsy. When Galton was among those duped by the Victorian craze for séances, Darwin could barely hide his contempt; "the Lord have mercy on us all, if we have to believe in such rubbish", Darwin wrote (Desmond & Moore 1991, p.608). And Galton did not stand in the way of the trade in slaves because it was a global system of terror, kidnapping, dehumanisation, torture, mutilation, rape and murder; he opposed the slave trade, as he put it in a letter to *The Times*, because it was inefficient, because it was "awful disorganization". There is a "lottery" to "slave catching", Galton wrote (1857). "I do not join in the belief that the African is our equal in brain or in heart; I do not think that the average negro cares for his liberty as much as an Englishman, or even as a serf-born Russian." Because of this, he continued, "we have an equal right to utilize them to our advantage. ... There can be no just complaint of tyranny. These persons are simply treated as children by their masters, and compelled to do what they dislike for their future good and for that of society at large". As their "masters" we can thus use any legitimate means, "or even *quasi*-legitimate means", to possess ourselves of their sweat and services, concluded Galton.

THE "DOUBLE DOSE OF UNFAIRNESS" ANIMAL

Far from Type II nature being as "*the* moral animal" (Wright, 1994), the "human decency is animal" animal (Wilson, 1975a) of the sociobiologists and even the group-selectionists, our biological inheritance is in fact to set us along an often irrational pathway that almost always ends in us becoming the "double dose of injustice" animal, the doubly indecent animal. The terms double dose of injustice and double dose of unfairness come from the writings of the Oxford anti-free will philosopher Neil Levy (2011, 2019). Yet just notice straight off that the developmental outcome of double unfairness is perfectly explained by abandoning, as did Darwin, belief in a single human inheritance mechanism, and accepting instead the uncertain (in a fully deterministic sense) interplay of two largely incompatible inheritance systems, as overseen by a limited capacity for reason that anyway has no end-point in and of itself.

In *Hard Luck: How Luck Undermines Free Will and Moral Responsibility*, Levy writes that Dan Dennett is just plain wrong to argue that luck averages out in human life, because luck "tends to ramify" (2011, p.199), both the bad and the good. "Chance events that are genuinely lucky and that actually compensate for constitutive luck are rare and extraordinary." If you start off with a poor biological – or environmental – hand it will largely follow you, particularly in countries like America and Britain that make a fetish out of the self-made-man myth. And that far from having evolved to be the fair and the just ape, most of us, conservative and liberal-left, are the doubly unfair ape, the doubly unjust ape. First there is the poor developmental luck many have to suffer while the more fortunate frequently look the other way. But free will justification then adds a *second* dose of injustice, Levy writes, onto that first dose of random indifference. This is the unfairness of claiming that the person was somehow responsible for their first dose of unfairness, or at the very least that blame and suffering are their just deserts. And free will apologists "double dip on reward", Levy points out, in that not only are they lucky enough to have benefitted from undeserved great good fortune but, as with the essentialist philosopher Gary Watson, will usually seek to claim that such good fortune was somehow due them, and even that they can take credit for it ("a matter of who we essentially were … this difference still might explain what is to my credit" – Watson 2004, p.248).

The predominant system, says Levy, amplifies good luck for the more fortunate, and amplifies bad luck for the less fortunate, while both denying that it is doing any such thing and making us all complicit in the unfairness and the injustice. Incapable of being fair or just, and further incapable of even fairly defining fair play and justice. We said above that the vile mindset of *sanchita* karma is used to legitimise discrimination against low castes through the creation of a cruel alternative reality where low castes are said to be paying for their past wrongdoings, and high castes are argued to be reaping the rewards of their past goodness. So it is worth highlighting that the *sanchita* karmic alternative reality really does seem to be the Eastern equivalent – morally, practically, and intellectually – of the Western alternative reality of free choice, and of Dawkins' "short-cutting" of a "truer analysis of what is going on in the world". We don't come close to being the moral animal of the sociobiologists, the biologically decent animal; most of us across the East and the West are the double dose of indecency animal, the double dose of immorality animal. What a low bar we have set ourselves, and yet we fail to clear even this.

Levy's double dose of unfairness point underscores that we are not necessarily even asking the more fortunate to assist the less fortunate. The undeservedly more fortunate, remember, and the undeservedly less fortunate, because dumb luck swallows everything, to use Galen Strawson's line. We are not asking the more fortunate to assist the less fortunate; we are not asking them to intercede in that first dose of unfairness. We are just asking them to tell the truth about that first dose of unfairness; we are just asking them to avoid piling on that second dose of unfairness. Yet we, including over 90% of humankind's most logical thinkers, can't even do as staggeringly little as that. What a low bar to fail at. A low bar that is not even open to opine over, because it is the low bar the logic of the universe forces us to confront. Our wretched failure, our woefully inadequate rational and ethical nature, is there for all to see at but a single glance. Intellectually and morally, we are so often like the other snow monkeys who made no attempt to assist the severely deformed Mozu "in her monumental struggle for existence". They behave as they do because they are Type I, pure nature, and cannot behave any other way. We behave as we do because we are Type II, nature plus nurture plus limited rationality, and we so often cling to ideologies that tell us not to behave any better. We are the double dose of unfairness ape, unwilling to be fair and further unwilling to even fairly define the concept of fairness, and to date without the wit or character to do anything about it apart from employ academics to make up more spin and shamefully false stories about how "luck averages out in the long run".

Yet this is what we can expect, probably all we should expect, of any Type II intelligence, anywhere in the universe, where reason is not an end in itself and simply operates under an easy-to– manipulate psychology. Of course they will not just have their Daniel C. Dennetts and their Steven Pinkers; they too will have their Neil Levys and their George C. Williamses. But the former will likely greatly outnumber the latter. And their societies will thus be structured as ours is, and this is of more than purely academic interest, because this then defines both the nature of a people and their attitude to opportunity. In 2018 Jim Ratcliffe, founder, chairman and majority owner of the oil, gas and chemicals giant Ineos, became the UK's wealthiest person, and by quite some margin, shortly before he was knighted, and then announced his relocation to the tax haven of Monaco. "But also, you know, I started life in a council house in Manchester. I've worked quite hard to get where I am today. … And you know that opportunity's open to everybody on the planet", Ratcliffe had previously told the BBC (Sackur, 2016). Ratcliffe went on to mention his admiration for America precisely because "America embraces … much more warmly than we tend to do in the UK" this understanding that taxation is to some extent unnecessary, as everyone gets the same chances to be wealthy. Or there is Amazon's Jeff Bezos, the world's richest person in that same year of 2018, giving the Baccalaureate Address at Princeton and telling the students that although their particular level of intelligence may not have been their free choice, was indeed a gift and not a choice, their personality, social position and economic success was their free choice. "Gifts are easy – they're given after all. Choices can be hard. You can seduce yourself with your gifts ... to the detriment of your choices. ... In the end, we are our choices" (Bezos, 2010).

That Type IIs are the double dose of unfairness animal is of more than purely academic interest, because what the above indicates is how little real understanding of the world the more fortunate often have. They lack appreciation of how – undeservedly – fortunate they were and are in life. They lack understanding of the absolute role luck plays in human life. And they lack understanding of how – undeservedly – unfortunate many others are in life, and that many have not the same opportunity to succeed, have not just less opportunity to succeed, but have zero opportunity to succeed. Because no one else starting life in a council house in Manchester had the opportunity to build an industrial powerhouse. It took a particular (undeserved) biology combined with a particular set of (undeserved) fortuitous environmental triggers, contacts, and life chances, to reach that singular outcome. And personality and success are not a matter of choice; they are ultimately as much out of your hands as the shape of your nose or the colour of your skin, which is why ongoing opportunity matters so very much. And why it mattered so much to Ratcliffe and Bezos when they were growing up, even if they have managed to conveniently convince themselves otherwise.

Writing for the *New Yorker*, Charles Duhigg commented that, "'Jeff is a libertarian,' a close acquaintance, who has known Bezos for decades, told me. 'He's donated money to support gay marriage and donated to defeat taxes because that's his basic outlook – the government shouldn't be in our bedrooms or our pocketbooks'". Amazon has an almost pathological aversion to paying taxes, and ended up with a serious publicity problem when it opposed, both through money and threats, a unanimously passed 2018 Seattle City Council measure attempting to address a spiralling homelessness crisis. The *New Yorker* continued: "Bezos's close acquaintance agrees: ... 'The one time Amazon could have pitched in, on the homelessness tax, instead of taking the lead Jeff threatened to leave. It's how he sees the world'" (Duhigg, 2019). It can clearly only be damaging for the world's richest and most powerful people to appear to buttress their social, economic and political views with palpable falsehoods, even if 90% of philosophers, and probably closer to 95% of the general public, are similarly blinkered and self-deluding. On any planet hosting a Type II intelligence, far from that Type II intelligence being "the" moral animal of their planet, that Type II intelligence will be the doubly unfair animal of their planet. Incapable of being fair and just, and further incapable of even fairly defining fair play and justice. With the opportunity to do better in the future, yes, though largely only once that Type II creature has summoned the courage to face its true nature.

TYPE II NATURE AND HUMAN NATURE

"The dominant male, Humphrey, held a struggling infant about 1.5 yr old. ... After 3 min, he began to eat flesh from the thighs of the infant, which then stopped struggling and calling" (Bygott 1972, p.410). All the orthodox biological evidence is that the sociobiologist Steven Pinker carries Humphrey's genetic code, the genetic code for cannibalism and infanticide. And Pinker appears to be descended from mindless religious fanatics who celebrated genocide, slavery, and the raping to death of children, while assiduously "avoiding intermarriage" with non-fanatics. Surely Steven Pinker is proof positive that even the most genetically benighted little sprog can have extraordinary social and economic success if just raised in a stable and supportive community, with access to significant opportunities, including tax-enabled educational opportunities. And isn't this understanding something we should want to welcome with open arms?

Yet on every planet with Type II intelligence there will be leading universities where professors who are the product of vast privilege and opportunity will deny they are the products of vast privilege and opportunity, and deny privilege and opportunity are necessary for others to have their degree of success in life. We know Pinker carries Humphrey's genetic code because the evidence is that we all do. Although almost as soon as Darwin in 1871 published *Descent of Man* and placed humankind firmly within the animal kingdom, the backlash of sociobiology started, even if it took a century for the name to settle. The backlash that seeks to uniquely lift human evolution above orthodox natural selection. Richard Owen would claim that humankind should stand in a sub-class "reserved for him alone", while Lord Kelvin would hold that humanity must have evolved under exclusive rules. And a century or so later, at Harvard, Bob Trivers and Steven Pinker would still be arguing on the basis of "no direct evidence … nor its genetic basis" that a unique "cognitive twist" of the rules of natural selection had reserved for us alone a creation called "fictive … faux-families", with "illusions of kinship". Darwinism is all about the struggle for existence. But all too often it seems to be about the struggle for existence between very different views on the evolution of our species. On the one side there are those, including Darwin and my friend George Williams, who try to place humankind within the orthodox rules of the natural world. To the other side are those, including Lord Kelvin and Robert Trivers, who wish to set humankind apart from the rules that apply to the rest of nature.

When George died, the *New York Times* commented that, "The importance of Dr. Williams's book was immediately recognized by evolutionary biologists, and his ideas reached a wider audience when they were described by Richard Dawkins in his book *The*

Selfish Gene. ... Dr. Williams acknowledged that people had moral instincts that overcome evil. But he had no patience with biologists who argue that these instincts could have been brought into being by natural selection" (Wade, 2010). Williams had no patience with these biologists because we know any Type II intelligence should carry the same biological code as Humphrey, Hide and Lukaja. We know this, because recall what it would have meant to have moved beyond Humphrey and Hide's genetic pattern in the six or so million years since we split from a common ancestor with chimps and bonobos. It would have required multiple vast leaps in genetic design space, multiple grand saltations. It would have required selection to have backtracked through the entire genome, deleting or inhibiting genes for common animal behaviours while substituting genes for antithetical behaviours never before seen in the history of evolution. It would have required the rewriting of a four-billion-year evolutionary pattern, the fundamental change to a billion-year gene– and individual-selection design that needed no change. It would have required the invention of new and wholly unexpected processes – "indirect reciprocity", "metamorphosed" phenotypic expression, "*mal*adaptive" evolution, or a "multiparty altruistic system in which altruistic acts are dispensed freely" – that are both evolutionarily unstable (at least according to the biologist who developed the mathematical concept of the evolutionarily stable strategy) and that we do not need in the explanation of all other life on Earth. It would have required the reworking of an outbred diploid organism into one with seemingly an inbred or haplodiploid behaviour pattern, but a behaviour pattern way beyond what nature had previously achieved even for an inbred, hard-wired, and reproductively suppressed organism. It would have required the infinitesimal 0.1% genetic distance between our "glorified chimpanzee" ancestors and ourselves to have been the locus of the drive for selfishness as old as life itself, or alternatively that we retained 99.9% of a cannibal's DNA yet "our modern environment" metamorphoses "beyond recognition" a billion-year behavioural archetype. Because the parsimonious answer is that we didn't lose anything in that final 0.1% change, and certainly not a four-billion-year pattern for pitiless indifference and pathological selfishness; we gained something. Just another cannibal ape gained a susceptibility to culture with the emergence of a much larger brain and the capacity for spoken complex language. Just another cannibal ape was now subject to two, largely incompatible, inheritance mechanisms.

Only Darwin's solution can explain how Honest Abe Lincoln, the Great Emancipator, could be simultaneously among the finest Americans of his generation and an ignorant white supremacist. "'I will also add to the remarks I have made … that I have never had the least apprehension that I or my friends would marry negroes if there was no law to keep them from it, [*laughter*] but as Judge Douglas and his friends seem to be in great apprehension that they might, if there were no law to keep them from it, [*roars of laughter*] I give him the most solemn pledge that I will to the very last stand by the law of this State, which forbids the marrying of white people with negroes' [*continued laughter and applause*]" (Lincoln, 1858). Only Darwin's solution can explain how his half-cousin Sir Francis Galton could be both an enthusiastic progressive within the British Association for the Advancement of Science and a world-class bigot. Africa should be given to the Chinese, Galton wrote in *The Times* in June 1873. While black Africans, he argued, were an "inferior" race of "lazy, palavering savages" unable to sustain on their own "the burden of any respectable form of civilization", we cannot give their continent to just anyone. "The Hindoo cannot fulfil the required conditions nearly so well as the Chinaman, for he is inferior to him in strength, industry, aptitude for saving, business habits, and prolific power". Yet "the Arab is little more than an eater up of other men's produce; he is a destroyer rather than a creator, and he is unprolific". Galton wanted Africa for the Chinese, albeit he viewed the Chinese as also displaying many character weaknesses; "the bad parts of his character, as his lying and servility" (Galton, 1873).

Because this vanity that we are, biologically, "*the*" moral animal explains nothing. It doesn't explain why half of America had to be forced at gunpoint to renounce first slavery and then segregation. This vanity doesn't explain a billion high or higher caste Indians, or a billion Muslims, or at least 200 million American Christians. It doesn't explain any of those people who willingly "surrender" critical thought for obedience without question, or the fundamentalist Russian Orthodox Church refusing to denounce murderous aggression and blatant war crimes. This vanity doesn't explain Abraham Lincoln's need to divide the world into superior and inferior races, or Galton's "Africa for the Chinese" thuggery, or billionaire businessmen irrationally believing that "everybody on the planet" has the same opportunities they had. This vanity doesn't explain the world's most influential scientists erroneously describing themselves as part of a genetic elect, or casually talking about "short-cutting a truer analysis" of what is going on in the world. And this vanity doesn't explain 90% of humanist philosophers; it doesn't explain nine-tenths of humankind's smartest and most educated thinkers.

"You will think me very conceited when I say I feel quite easy about the ultimate success of my views, (with much error, as yet unseen by me, to be no doubt eliminated); & I feel this confidence, because I find so many ... can thus group & understand many scattered facts", Darwin wrote to his former friend Sir John Herschel in May 1861 (Warner 2009, p.438). It is because of Darwin we can today group and understand many seemingly scattered facts. From cosmic dinosaurs – because transition four is not the main bottleneck – to a lack of flying dragons yet a fossil record of dragonflies with wingspans over 70 centimetres. From sharks with a bite force ten times as powerful as a great white to

Mozu's solitary and "monumental struggle for existence". From the behaviour of E.T.'s leathery compatriots to the nature of James Cameron's *Alien*. From the mindless and genocidal fanaticism of our religious ancestors to artificial intelligence that will exterminate us not because we are ants but for the sole reason that it remains the ant. It is because of Darwin that we can understand both Type I nature and Type II nature, anywhere in the universe.

8

THE THREE FLAVOURS OF TYPE II EXISTENCE

> "It seems probable that once the machine thinking method had started, it would not take long to outstrip our feeble powers. ... At some stage therefore we should have to expect the machines to take control."
>
> – **Alan Turing**, mathematician and father of computing (1951, p.109)

The first part of this chapter is necessary in order to tie up loose ends. After all, how did we get to the current state of affairs? Darwin's explanation of higher intelligence, and the gene-selectionist explanation of higher intelligence, would be a fundamental shift away from our current interpretations. Higher intelligence, for Darwin, is simply impossible with one natural inheritance mechanism, and requires two inheritance mechanisms, and in fact two often warring inheritance mechanisms. But intelligence is just part of this new understanding, because for Darwin, intelligence, large group harmony – what group-selectionists like Martin Nowak even go so far as to call "superCooperation" – and morality are an inseparable triad, a transitional troika. So if Darwin and the gene-selectionists are offering us such a radically different interpretation, how did we get to a situation where this is not being openly discussed and evaluated? There are three different evolutionary interpretations of the origins of intelligence, morality and harmonious co-operation, being gene-selectionist, group-selectionist and sociobiologist. If the three interpretations were equally likely to be giving the correct answers it would be worrying enough if any one of the three was not being properly evaluated. But if what seems to be the most likely answer, Darwin's answer, is the one that no one will discuss? Darwin's answer that has profound implications not just for non-terrestrial intelligence, but for self-aware artificial intelligence? Then again, the two traditions that we *are* prepared to discuss have similar profound implications for non-terrestrial intelligence and self-aware artificial intelligence that are *also* not being investigated. So how did we even get to this situation? This chapter will seek to tie up such loose ends, but by doing so it will further allow us to see in detail what underlies the three "flavours" of Type II biology, being the three different models presented by Darwin and those convinced that Darwin got it wrong on human evolution.

CASUALITIES OF WAR

"Maynard Smith, Williams, Hamilton, and Dawkins ... have largely eschewed the deeply unpleasant task of pointing out more egregious sins in the work of those who enthusiastically misuse their own good work", the philosopher of science Dan Dennett wrote in his bestselling book *Darwin's Dangerous Idea* (1995, p.485). What is Dennett referring to here? What are these "sins" that biologists have turned a blind eye to, and indeed can we really have had a situation of professional biologists seemingly tolerating decades of misrepresentation of orthodox Darwinian theory?

Readers thus need to appreciate the history, and appreciate that those enthusiastically misusing the good work of Maynard Smith, Williams and Hamilton came to have undue influence in the late twentieth century because they were at first being used by biologists like Richard Dawkins, and even John Maynard Smith, who then lost control of them.

During the final quarter of the twentieth century there was an open war taking place within evolutionary biology. Much of the emotion was over levels of selection, and was partly a reaction to the runaway success of Williams' and Maynard Smith's genic selection theory, but there were other differences too. For instance, Darwin's evolutionary gradualism seemed to be under further attack from Stephen Jay Gould and Niles Eldredge's theory of punctuated equilibrium, which invoked long periods of stasis broken up by rapid and only occasional change. Please do try to recognise the significant damage that the leading gene-selectionists feared could be done to evolutionary orthodoxy by a resurgent group selection, or even just the distraction of punctuated equilibrium. Darwinism had only just emerged from a very long and somewhat confused period in which vague Kropotkin-like suggestions about nature selecting at the level of the group had eventually culminated in V.C. Wynne-Edwards' 1962 claim that nature could and did select for the benefit of the group even at the expense of the individual. Williams and Maynard Smith's genic selection theory had swept across academic biology because it put an end to confusion and vague references, and used rigorous mathematics to test evolutionary stability, yet America still seemed to be lagging behind, and to some threatening to reverse the gains already made.

As Richard Dawkins put it in the 1989 second edition of *The Selfish Gene*, while group selection was out of favour with British and European naturalists from the early 1970s this was most certainly not the case with large parts of American naturalism.

Group selection, he wrote, "is even more out of favour among biologists than it was when my first edition was published. You could be forgiven for thinking the opposite: a generation has grown up, especially in America, that scatters the name 'group selection' around like confetti" (1989, p.297). It is littered over all kinds of cases, Dawkins noted, that used to be, "and by the rest of us still are", clearly and straightforwardly understood as something else, such as Hamiltonian kin selection. "It is irritating to find that we are now two generations, as well as two nations, divided only by a common language." At least in the popular press the war was regularly presented as Dawkins versus Gould, although even the ever-thoughtful John Maynard Smith and the gentle George Williams would end up saying some unwise and unnecessary things. But the worst mistake gene-selectionism made was that it started a proxy war. It allowed the later incarnation of sociobiologists to wade in on its behalf, and speak in its name. It did this as it hoped to free up the considerable time it was wasting fighting distracting battles but also because, frankly, it took too much pleasure from seeing group-selectionists, and particularly Gould, Eldredge and the geneticist Richard Lewontin, savaged. Though the leading gene-selectionists feared the damage that the sociobiologists could do to Darwinism, and most obviously their suggestion that humankind had broken from the orthodox rules of genic selection, they feared (re)losing ground to the group– and multilevel-selectionists even more.

The journalist Andrew Brown catalogued "the Darwin wars" in the 1990s, and he gives us his view of the character of the later sociobiologists. Gould was particularly disliked by them, Brown wrote, reporting that Leda Cosmides and John Tooby's attack on Gould was "so blistering" that the *New York Review of Books* "refused to print it" (1999, p.150). Brown's position was that their attack was "extraordinarily ... unjustified" against "a man who has won just about every prize going for science writing" (p.151). Gould had previously fought successfully against abdominal cancer – cancer that he likely developed from asbestos installed in the 1930s within the structure of his comparative zoology museum at Harvard – and for an example of what he was up against before his early death from lung cancer in May 2002 see the sociobiologist Robert Wright's December 1999 piece "The Accidental Creationist: Why Stephen Jay Gould is Bad for Evolution". Wright used his attack on Gould to proselytise for sociobiology, telling us not only that "some noted biologists, such as William D. Hamilton and Edward O. Wilson, believe that the evolution of great intelligence was likely from the start", but that "Hamilton's work also suggests another interesting likelihood … that evolution's directionality may have a 'moral' dimension". A suggestion being made, remember, with "no direct evidence … nor its genetic basis", and reversing both the rules and mechanisms nature had exclusively used for the preceding billion years. An image of pots and kettles comes to mind.

Gould could certainly be exasperating and distracting, but he probably did more to promote the teaching of evolution in the schools of the United States than any other scientist. Though another problem, partly driving the more right-wing sociobiologists' personal loathing, was the pretty obvious left-wing ideology of many of the leading group-selectionists. Gould and Lewontin were key members of what was called the Sociobiology Study Group, an academic organisation formed to counter sociobiological explanations of human behaviour, and which later associated itself with the Boston arm of Science for the People, thereby becoming the larger Sociobiology Study Group of Science for the People. Science for the People was, however, an avowedly left-wing organisation that had grown out of protests against the Vietnam War.

The relentless left-wing political attacks on (and unfortunately then conflating) both genic selection and sociobiology ended up incensing respected biologists from all backgrounds. Historian of science Ullica Segerstrale writes that Ernst Mayr, key thinker on speciation and one of the foremost contributors to the modern synthesis of Darwin's work with the genetics of Gregor Mendel in the early and mid twentieth century – and co-recipient with Williams and Maynard Smith of the 1999 Crafoord Prize – admitted to her that it would have been "so easy" to criticise sociobiology on purely scientific grounds. "Perhaps even more damaging to the potential scientific debate about sociobiology was the absence of any serious critical reviews in scholarly journals. There was a clear reason for this." According to Mayr, writes Segerstrale, just because of the political criticism "several people who had been severely critical of sociobiology and had taken their time preparing reviews for scientific journals, now simply tore them up" (2000, p.17). Segerstrale's point was these highly critical commentators withdrew their denunciations as they did not want their "genuinely scientific disagreements" to be seen as in any way supportive of the Sociobiology Study Group's political attacks on sociobiology; "Mayr knew of at least three such cases". Segerstrale then went on to draw a parallel with other examples from the history of science where few take on the real responsibility to correct others' mistakes. "And fewer still regard it as their task to enter the controversy in order to increase mutual understanding between parties in a polarized situation. It seems that most scientists are content with taking a passive stance, from which they watch the show while they get on with their *own* research" (p.245).

To give two examples of actual events on the ground during this period, firstly we have Segerstrale writing that when left-wing critic of gene-selectionism the biologist Steven Rose ("Britain's Lewontin" – p.7) reportedly threatened in 1985 "to sue Dawkins for libel … there was quite a flurry of activity at the time to protect Dawkins" (pp.191–2). Bill Hamilton approached E.O. Wilson in order to get Segerstrale's PhD thesis for use in a possible legal defence, and Dawkins himself contacted Segerstrale for background information. The second instance shows how the Dawkins versus Gould war also impacted other disciplines. The Cambridge palaeontologist Simon Conway Morris produced an influential book in the late 1990s analysing and interpreting the fossils from the famous Burgess Shale in British Columbia. In *The Crucible of Creation* Conway Morris had savaged Stephen Jay Gould's account of the Burgess Shale investigations and the supposed Cambrian explosion of life. Savaged may be the right term here; Richard Fortey, a senior palaeontologist at the Natural History Museum in London, wrote in his book on trilobites that he had "never encountered" such hostility "in a book by a professional; I was taken aback" (2000, p.136). At least according to Fortey the rancour was undeserved by Gould who, he claimed, had been nothing but honourable in his treatment of Conway Morris. But what is germane here was Dawkins' subsequent lauding of Conway Morris – a devout Christian, so not normally to be defended by Dawkins – simply because he was attacking Gould. Fortey, a Cambrian era expert, certainly appeared to believe that some of those, like Richard Dawkins, who drew attention to Conway Morris's criticisms of Gould, may not have been fully grounded in the history of the Cambrian explosion-of-life opinions. Opponents of Gould in other arenas, Fortey wrote, they appear to have used the book as a stick to beat him, "operating on the principle: 'my enemy's enemy is my friend'" (p.138).

The above may explain some of the betrayal of Darwin, but not all, and now history will develop along a different path. Gene-selectionism had kept sociobiology around to fight its battles for it, but in doing so it made a critical error. Because Maynard Smith's "son of sociobiology", the later incarnation of sociobiology, had been relabelled as evolutionary psychology, and thus as a form of psychology, it could thereafter only be performed by social scientists; by "psychologists, anthropologists, economists, historians" (Horgan 1995, p.151). Psychology is a soft or social science, not a hard science like the natural and the physical sciences, and rightly or wrongly there is seen to be a scientific pecking order of hard to soft driven by capacity for objective observation and experimentation, something often impossible in the social sciences. A biologist will not generally write within a field that incorporates the term "psychology" in its title, in part because of inherent weaknesses within that discipline, including well-known problems with its methodological competence and statistical understanding. As a *Nature* headline put it in August 2015, "Over Half of Psychology Studies Fail Reproducibility Test", while the article itself went on to say "don't trust everything you read in the psychology literature. In fact, two thirds of it should probably be distrusted" (Baker, 2015). Yet by abdicating human evolution across to the psychologists and social scientists, biologists had let themselves be frozen out from investigating what Darwin himself had termed "the highest & most interesting problem" for the naturalist, and from what Wallace had called "the inverse problem" for the biologist. The most interesting problem in evolutionary biology – a puzzle with vast implications for multicellular life off-planet, a puzzle over Type II take-off with consequences for all biological intelligence anywhere, alongside questions over large group harmony, behaviour and morality, and a puzzle with enormous implications for our current dabbling with artificial intelligence – and biologists had stood themselves down from all these debates. Maynard Smith excused this abdication of responsibility by noting that "human societies change far too rapidly for the differences between them to be accounted for by genetic differences between their members", and "as differences are what we are primarily interested in, there is little an evolutionary biologist can say" (1992, p.82). At least according to this statement, evolutionary biologists actually have no duty to correct psychologists and other social scientists who blatantly misrepresent evolutionary orthodoxy and advance misplaced biological solutions for the human animal.

But it was also hostility from outside biology that had created what Dennett in *Darwin's Dangerous Idea* called a "siege mentality" (1995, p.485) within orthodox gene-selectionism that had led the leading theorists to eschew their "duty" to expose the sins of the sociobiologists. There is generally little love lost between scientists and philosophers, with a number of the gene-selectionist biologists having seen themselves as being particularly provoked. In 1979 the late philosopher Mary Midgley wrote a paper for the journal *Philosophy* criticising, among other things, Dawkins' use of metaphor, a device he often uses to get a larger message across. The paper was quite extraordinarily unfriendly; *Philosophy* permitted Dawkins a reply, and his bitterness was manifest. "Such transparent spite ... so rude ... I deplore bad manners ... it is hard for me not to regard the gloves as off" (1981, p.556). Midgley, wrote Dawkins, raises the art of misunderstanding to dizzying heights, and had "so pathetically misunderstood" the mechanism of reciprocal altruism (p.571). But in his renunciation of such philosophical criticism, Dawkins was quick to defend sociobiology and "our field (it wasn't called sociobiology then)". Time did not heal these wounds and, according to the journalist Andrew Brown, in 1992 Dawkins was still so offended by Midgley's review of his work thirteen years before that he withdrew from a conference to which he had been invited "when he heard that Midgley would also be present. 'I wouldn't want to see her over breakfast,' he told the organiser" (1999, p.87).

Returning to the point *Nature* was making in 2015, and that two-thirds of what you read in psychology – evolutionary or otherwise – should probably be distrusted. In 2012 the Nobel Prize– winning psychologist and economist Daniel Kahneman had warned his fellow psychologists that there was a "train wreck looming" for the field. As *Nature* then put it, "Nobel Laureate

Challenges Psychologists to Clean Up Their Act" (Yong, 2012), after Kahneman voiced concerns that some of his colleagues were not robustly challenging their own evidence. Kahneman's concerns were part of a wider criticism of academic psychology. In 2015 psychology was rocked to its foundations when the Center for Open Science-led "Reproducibility Project", a collaboration of 270 contributing authors trying to repeat almost one hundred published experimental and correlational studies, failed to verify many of those key social psychology papers, and hence the *Nature* article warning over psychology studies. Part of the background to this is that the social sciences, and particularly large parts of psychology, have in the past earned a reputation for not understanding statistical relevance, at least according to statisticians.

So, for example, a 2016 article by the Turing Fellow, theoretical physicist, and Oxford data scientist Taha Yasseri is entitled "P-values Are Widely Used in the Social Sciences, But Often Misunderstood: And That's a Problem". *P*-values, probability values, are a measure of statistical significance, but what is called "*p*-hacking" is a common worry in the social sciences, as getting a desired result is not difficult if data are improperly sorted. A 2011 survey of over 2,000 psychologists at major US universities found that the percentage of respondents who had engaged in "questionable practices was surprisingly high", and where some questionable practices may even be "the prevailing research norm" (John *et al.* 2012, p.524). So great are statisticians' concerns over *p*-values – some now refer to "*p*-trash", and "*p*-dolatry" – that in March 2016 the American Statistical Association released a statement on the statistical significance of *p*-values "intended to steer research into a 'post $p<0.05$ era'". Although a *p*-value of less than 0.05 does not mean that there is less than a 5% probability the experimental result is due to chance, surveys show that most social scientists appear to mistakenly think it does mean that. Hence the ASA statement actually had to spell out to the social science community that "*P*-values do not measure the probability that the studied hypothesis is true, or the probability that the data were produced by random chance alone" (ASA, 2016). Many of the hard sciences work to much higher statistical significance tests. In particle physics, for example, the 2012 discovery of the Higgs boson satisfied a test of five-sigma, or five standard deviations. High-energy physics tests to five-sigma because a number of three sigma events have later turned out to be nothing more than statistical anomalies, background noise. Yet many psychologists and social scientists continue to resist calls for more robust analysis of their work, partly fuelling the hard science community desire to maintain separation from these fields.

BEHAVIOURAL GENETICS – AN EPISTEMOLOGICAL REVIEW

Given this hopefully interesting historical and methodological background, we must now ask a crucial question: Did Darwin get intelligence, large group cohesion, and morality wrong? Because there are empirical studies out there that claim the sociobiologists got it right and that Darwin got it completely wrong, and that humans, or at least most humans, escaped from the billion-year template to evolve to be naturally moral. Other related studies then go on to claim that success and social position are largely genetic (that benign biology "predicts a variety of positive life outcomes" – Pinker, 2006), and even that affirmative action and government spending may be wasted ("an injustice" – Sullivan, 2011a) on what may be a perfectly natural and ineradicable economic and social underclass, even an innate colour-based intellectual underclass. It is time to consider the field of behavioural genetics.

We will put to one side for a moment the question about the extent to which such studies might be methodologically competent. We will also put to one side how such studies can claim to even show that the vast majority of contemporary humans are born naturally moral when for most of human history – including the great majority of American history – discrimination and a double dose of unfairness has been the order of the day. We will not yet ask how these studies can be trying to deconstruct moral cognition within a species where almost all remain woefully ignorant of, and indeed largely act contrary to, the single moral fact humanity can ever know. (Note here that philosophers like Dennett and Smilansky may not want us to pull back the curtain and meet the wizard, but they have at least worked out where the wizard is; the great majority of behavioural geneticists haven't so much as noticed the curtain.) We will also delay asking how the conservative centre is being argued to be born naturally moral when almost all are descended from fanatical ancestors whose atrocities "lit up hell-fires". We will put these potentially valid epistemological concerns aside for now, and initially just ask whether it might be possible to have *both* Darwin being correct that morality is not natural, *and* at the same time have studies that might seem to show, at least from some angles and interpretations, that there is a biological spectrum to moral behaviour. Can we in effect have a two-stage developmental process that has led many to incorrectly conclude that we have only a single-stage developmental process?

Start-point	Type II Filter	Baseline	Genetic difference	End-point
Amoral ape	Culture / language	Culturally-affected human	1. Baseline 2. Impulsivity	1. Success 2. Less success

Consider the above table, and where we are moving left to right. We still start at the very left of the table with the billion-year pattern, the billion-species pattern of orthodox natural selection, of "no evil and no good, nothing but blind, pitiless indifference". There is then the Type II filter of language mediated through a brain susceptible to hopes and fears which *resets* the effective start-point for *all of us*. Because we have two inheritance mechanisms a naturally amoral ape has now become a culturally-overwritten human. But this will still leave culturally-overwritten individuals with different genetic loadings outside of the key attributes culture has deliberately over-written. So someone with natural impulsivity may have more difficulty concentrating, more of a struggle at school, and this goes on to lead to less success in adult life, including less success in the workspace. Or someone who is naturally slightly smarter then goes on to better economic success.

But notice something extraordinary here, something that seems to undermine much if not all of the academic discipline of behavioural genetics. Many of the studies alluded to above try to show that because some may have naturally lower IQ, or a degree of natural impulsivity, or more innate anger, that "this fact has social significance". *But it only has social significance if you ignore the first three columns.* Steven Pinker and Andrew Sullivan both try to claim that their undoubted economic and professional success was down to natural intellectual superiority. Now maybe in part it was, but that is no longer really relevant. Both Pinker and Sullivan owe their current success 100% to culture. The "natural" Steven Pinker, the Steven Pinker that began life on the very left of the above table, would be fit only to live in a cage, and as a cautionary tale zoo exhibit; it required the cultural "beating out of his nature" (to co-opt that wonderful line from George Price's biographer) to even begin to allow Pinker to have professional, social and economic success. It required the targeted application of the necessary tax-funded social and educational resources and opportunities, which is also all affirmative action is, to hand Pinker his success on a plate. Thus it now becomes almost totally irrelevant that Pinker's natural smartness allowed him to become a professor at Harvard; differential biology suddenly becomes largely meaningless.

There is still a residual biological difference, but in effect all we are now saying is it might have required sixty hypothetical social investment "units" to move Pinker from cannibal ape to Harvard professor, while it might then take seventy such units to move another from cannibal ape to low-wage job. Yet Pinker's success is nevertheless 100% down to cultural overwriting, and he got exactly the cultural conditioning required to hand him his success. The natural Steven Pinker would have been incapable of holding down just about any job beyond raw meat taster. So how can anyone suggest that it is at all meaningful ("this fact has social significance") to note a distinction that one needed sixty units, yet the other is supposed to be inferior and a burden for requiring seventy units? They both wholly required the same fundamental beating-out of their natures, even if one took a little longer or cost a bit more. So how can anyone morally or socially differentiate, or go on to suggest, for example, that the "right" cut-off for society is sixty units of overwriting? Why not fifty-five units? But if we'd cut off at fifty-five units, the brilliant John Maynard Smith might still have been fine but maybe Pinker would then have ended up as our feral problem, to be chased after with a net? Though human development is so messy it will always be next to impossible to realistically argue how many comparative "units" any person actually required, including that Pinker might have needed only sixty units while someone else might need a higher level of seventy units. Though let us at least note for the record that the father of behavioural genetics and modern eugenics, Sir Francis Galton, came from a staggeringly privileged background, and thus at least on the face of it received far more than the above "burden" number equivalent of seventy social investment units, and still ended up a fool and a snob,[5] and never came close to being the moral animal. Galton personifies what might be called the paradox of eugenics; if there actually were to be anything real behind the eugenic propaganda, eugenicists tend to be the last people we should logically want to breed from.

Many will still be trying to get their heads around the above table, so let us try for an analogy. Think of the artificially selected domestic dogs from the third chapter, recalling their hypersociability and the extension of juvenile behaviours into adulthood. Then jump back to Steve Jones' dual comments that, at least for Darwin and the gene-selectionists, there is no charity in biological nature, and that we welcome into our home a "beast that preserves much of its primordial self. Overgrown juveniles though they are, evolution by human choice has not removed the instincts of their ancestors". Similarly cultural "domestication" has not removed our ape instincts for cannibalism and infanticide, but it leaves us hypersocial, and arguably even a tad juvenile in our mental outlook. Yet remember dogs can still have different genetic proclivities, even if all are domesticated and therefore all fit to live with us and in our homes; some are naturally more friendly and impulsive, some naturally more aggressive, some naturally smarter. But these biologically domesticated animals still retain their Type I genetic behaviours, just as we culturally domesticated Type IIs do. "Like wolves, dogs attack the weak … The homicidal packs relive their past." All dogs require a good home and good treatment to avoid the risk of them becoming feral, and needing to be chased after with a net. And the same is true for humans; all humans. This of course remains a somewhat inadequate analogy, because there

---

5   Galton thought England had "got rid of a great deal of refuse, through means of emigration" to North America. "As a rule, the very ablest men are strongly disinclined to emigrate", because the talented "prefer to live in the high intellectual and moral atmosphere of the more intelligent circles of English society, to a self-banishment among people of altogether lower grades of mind and interests" (1892, pp.360–1).

can be no possible analogy for what goes on in the human brain, as there is only one Type II species on this planet. The human experience has neither involved the intense and relentlessly one-way artificial selection pressures on our genome, nor has any correlate to the variety of breeds we have sought in animal domestication. We have performed artificial selection for the domestication of dozens of animal species, but the effects of cultural domestication stand on their own. Nevertheless, and perhaps because it is likely to offend and distress, the analogy with animal domestication remains important, if for no other reason than we should let it stand until we find a better way to discuss what has been done to us, and what we do to ourselves, through culture. We are unique, at least as far as we know today, a billion-to-one change to the way things have always operated, but the analogy will help somewhat illuminate a process – the move through transition eight – that we currently do not openly talk about or even intellectualise.

At least according to Darwin, every human being retains Type I biology but goes through the same Type II filter of cultural conditioning; we cannot avoid it, we are raised within the human language and concepts milieu that is central to taking an ape and making a human. There have been stories of feral children raised by wolves or such like, but these stories turn out to be false, or part– truths about abandoned older children and runaways. There is also the – one can only hope apocryphal – tale of the Holy Roman Emperor Frederick II in the thirteenth century raising two infants in complete silence and without human interaction, to see whether they naturally grew to speak a language imparted by the Christian God. They did not, and anyway it was impossible to raise them without some form of human communication. So it appears that we can reconcile Darwin with what may still be coherent reports from behavioural genetics, even if such studies are epistemologically highly questionable by failing to comprehend what we called in the above table both the genetic "start-point" and the later "baseline". Under the above model of potential Darwin-Galton reconciliation, behavioural genetics stops being *wrong* and become merely *unimportant*. But even with this theoretically being the case, the huge additional problem with behavioural genetics is that so many of the reports appear to be deeply flawed, fuelled often by truly extraordinary credulity, ideology and money, and serious methodological weaknesses.

"Professor Thomas J. Bouchard was sitting in his office at the University of Minnesota when one of his graduate students came in with the *Minneapolis Tribune*. 'Did you see this fascinating story about these twins who were reared apart? You really ought to study these.' Bouchard began to read the story. ... Bouchard thought it was odd enough that both were named James, but it was uncanny that each man had married and divorced a woman named Linda, then married a woman named Betty" (Wright, 1997a). This article comes from the London *Times* newspaper, and Bouchard has gone on to become one of the most influential behavioural geneticists of the last half-century, but just read this piece again. And then recall Dobzhansky's famous dictum: Nothing in biology makes sense except in the light of evolution. You see, the above *Times* article is complete and utter garbage from an evolutionary point of view, and doesn't make any sense biologically.

Posit if you so wish, and as sociobiologists often do, that genetics makes us look for partners who are tall, or attractive, or rich. That is evolutionarily at least plausible, or certainly worthy of investigation. But to suggest that genetics makes us look for partners called Billy-Bob or Betty is, frankly, bonkers. Bouchard was reported as suggesting that it was both academically and scientifically respectable to consider that the twins might be programmed to choose these women because of their names, as otherwise bringing up their names has no bearing ("within an hour of reading the article, Bouchard excitedly persuaded university officials to provide some grant money to study the

Jim twins"). But since none of our character attributes has any connection whatsoever to our externally imposed names, we get to Dobzhansky's point that there is simply no way for such a gene to have been selected. Unfortunately, though, most behavioural geneticists are completely unaware of the mathematics of evolution, though nothing in biology makes sense unless you can explain it evolutionarily. Yet behavioural genetics is not a branch of evolutionary biology. It is not a branch of molecular biology. It is not a branch of lab-based genetics. It is not even a hard science. It is a soft science, it is a branch of psychology ("twin studies are the very foundation of this branch of psychology" – Wright 1997, pp.44–5), that academic discipline currently going through a crisis of confidence because so many major studies have now been shown by those outside the discipline to be incapable of replication, that discipline that often profoundly misuses or misunderstands *p*-values and statistical relevance.

This anecdote has been repeated in so many influential places. Lawrence Wright is a Pulitzer Prize-winning American journalist, and his book *The Times* took this excerpt from was shortlisted for the Royal Society's annual science book prize. Steven Pinker has enthusiastically drawn attention to the "spooky similarities" (1994, p.327) that Bouchard has come up with. However, perhaps the most shameful instance of all was when the BBC allowed Professor Lord Robert Winston to screen a three-part documentary in July 1999 called *The Secret Life of Twins* around the "uncanny" similarities of identical twins. Baron Winston, a Labour life peer, has been the trusted go-to face of UK medical broadcasting for the last three decades. He is emeritus professor of fertility medicine and a professor of "science and society" at prestigious Imperial College London, is both Founder and Chair of the Genetics Research Trust at Imperial, and has been awarded the Faraday Prize by the Royal Society for excellence in science communication. Yet Winston's "major new BBC series" was actually advertised using the wives' story of the Jim twins, as they are known on the interview circuit. You can still find Winston talking to the Jims on Youtube (StuChannel3, 2010). This interview was included in the trailer the BBC ran with, and where one comes on screen to announce: "my first wife's name was Linda", then cut to the other announcing: "my first wife's name was Linda", then Jim I comes back for "I divorced Linda and married Betty", tick for Jim II, then "my first son's name was James Alan", and again tick the box for Jim II. Let me remind, this whole exchange is complete and utter nonsense for anyone actually interested in genetic influence; Dobzhansky's statistical fluke presented as real science by either the cynical or the bizarrely credulous. Just imagine all those conversations that took place around the office water cooler the next day, about how "spooky" it was that identical twins, and presumably the rest of us, may be genetically programmed to choose spouses by their names, after this major new BBC series as presented by a trusted science communicator quite disgracefully told the British public that these coincidences were scientifically significant.

Twin studies, and behaviour genetics more widely, speak to our desperate need to believe. It was interesting that Winston felt it necessary to interview Bouchard about his reported finding that religiosity is strongly genetically influenced. Winston is very unusual among British scientists in that he is deeply religious, and in fact combatively religious, and Bouchard's ostensible validation of Winston's "nature" seemed important to him. Yet the lack of scientific rigour here is concerning. Science is done by setting up control groups and comparing experimental results against the control. Yet many identical twins both reared together and reared apart but later reunited define themselves by their relationship with one another, much more so than non-identical twins do, and partly because of the attention factor. For example, identical twins take over an entire town in America each August ("Twinsburg" in Ohio) and have since 1976, and hold festivals, and give each other awards.

Yet there is a serious question mark over the value of studying two subjects who have already had a chance to interact, and a chance to demonstrate "odd" and "uncanny" similarities to the *Minneapolis Tribune* before you even get to interview them. Furthermore, some experiments become impossible to replicate, when replication is the cornerstone of scientific method, because for example there is a severely limited pool of monozygotic twins reared apart and later reunited. Such social engineering is no longer performed or permitted, and the problems of inadvertent coaching or fame-seeking make it dangerous to reuse subjects already interviewed.

The psychologist Steven Pinker has argued that notwithstanding a quite appalling history of doctored and often openly racist and chauvinistic studies published across behavioural genetics, including decades of financial support and politicised promotion from the American hard right, we can at least today trust the field, because in the mid 1990s "the American Psychological Association commissioned an ideologically and racially diverse panel of scientists to review the evidence" (2006). Okay, so this is that same APA that failed to see Kahneman's "train wreck" coming? The same APA that kept describing how robust the key published research was in social psychology, until hundreds of statisticians blew apart this conceit in 2015? The same APA that did nothing, even though many (Waller 2011, 2015; Zwaan 2013; Miles 2013) had for years been carefully explaining the intellectual and methodological weaknesses in published psychology studies promoted by the APA? Yet third-party oversight of behavioural genetics has been, if anything, worse. Following up the APA study, and indeed feeling the need to follow up the APA study with one not commissioned by the field actually being studied, in late 2000 the UK's Nuffield Council on Bioethics announced that it had set up a working party to establish the veracity of claims made in behavioural genetics. The Nuffield Council spent two years investigating the field before ... then asking a behavioural geneticist to write their main conclusion for them. To quote: "A review of research into [the genetic basis of] antisocial behaviour was written by a member of the Working Party Professor Terrie Moffitt [and this paper was] used to inform the Working Party" (2002, p.193). This point is raised not as in any way a criticism of Moffitt or her work across behavioural genetics, but because of the absurdity of spending two years investigating the veracity of a controversial field, only to ask a leading exponent of the area being investigated to provide the main conclusion.

Far more robust challenge to the psychologists behind behavioural genetics appears to be needed from the statistical and hard science communities. Because we can reconcile Darwin with what may be coherent reports from behavioural genetics, by differentiating between what we have called the genetic "start-point" and the later human "baseline". The Type II filter of language mediated through a brain susceptible to hopes and fears resets the baseline for all of us. So studies within the field of behavioural genetics can at one and the same time be correct yet have little to no social significance, as behavioural genetics is misunderstanding, is actually blind to, the first three tabulated columns above. Social psychology was finally audited in 2015, with stark lessons for the whole discipline, yet behavioural genetics, potentially a far more divisive branch of psychology even if done properly, has never been independently audited. The jury is therefore currently out on how much of behavioural genetics is legitimate but badly reasoned, and how much is just plain wrong, being part of that educational "train wreck" that is still so much of modern statistical psychology.

THE THREE "FLAVOURS" OF E.T. TYPE II

This review of behavioural genetics finally finishes our investigations, and allows us to complete the three alternative models of Type II intelligence and Type II coexistence, as presented

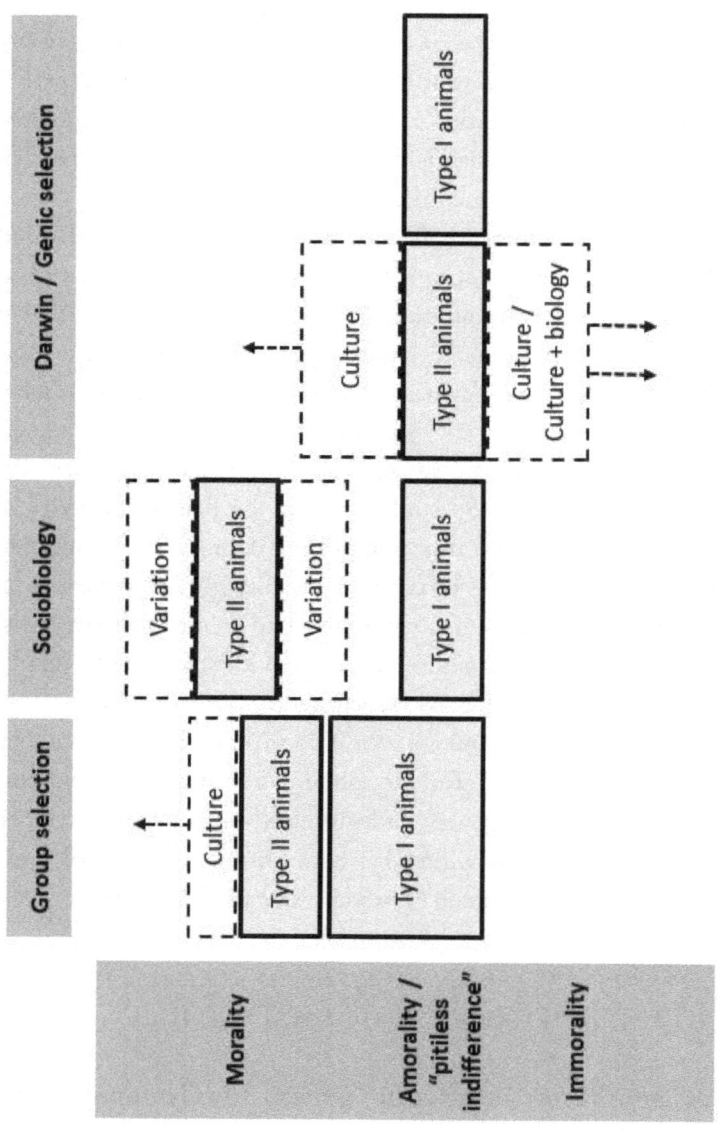

in the diagram on the opposite page. Remember that only one of these three traditions can be right, but that one of the three *will almost certainly have to be* right. And although these are three very different interpretations of evolution, just to remind that all three still give us cosmic dinosaurs and genocidal AI, because all three traditions recognise something very unusual happening beyond transition six, the evolution of multicellularity, and outside of transition seven, the evolution of eusociality and colonial living. And it is this very, very unusual occurrence beyond transition six and transition seven – an occurrence that only happened in one species in hundreds of millions of years of animal evolution, and in only one in a billion species – that is therefore almost certainly the greatest bottleneck in evolution, and thus the *prima facie* answer to the Great Silence. For all three of these alternative models, the Great Silence comes from out of a cosmic background noise of screeches, howls and roars.

We must thus summarise what each tradition is offering us beyond just dinosaurs everywhere and genocidal machine intelligence. Turning to our first model, the group-selectionists accept the amorality of Type I animal life, accept Humphrey and Ntologi's individual selection genetic code, but argue that when it came to humankind nature changed the operation of natural selection. Humankind evolved to be a radically different form of animal life, the Type II, in the most important aspects built under the rules of genetic group selection. Type II out-evolves and even reverses Ntologi's behavioural genetic code and stands apart from the blind pitiless indifference that chimpanzees and seemingly all other species must ultimately answer to. For the group-selectionists, Type II is the only creature built largely under the rules of genetic group selection, though some group-selectionists argue that even in Type I species, rare and occasional animal characteristics, such as bonobo adoption of out-group members, might also have a group-selectionist component, even if bonobo cannibalism does not. For the group-selectionists, we have dinosaurs everywhere, we have pitilessly indifferent Type I extraterrestrials as a relative commonplace as brain size has been de-linked from that further evolutionary transition, but we also have at least one animal species out of a billion that breaks the normal rules and evolves to be moral. The group-selectionists offer us the possibility of a truly wise, virtuous and avuncular E.T., albeit as a very rare evolutionary prospect, and so long as we accept the possibility of selection operating with major effects at the level of the group.

Moving on to our second model, the sociobiologists accept the amorality of Type I animal life, accept Humphrey and Hide's genic selection biological code, but argue that when it came to humankind, nature changed the rules of natural selection. Humankind evolved to be a radically different form of animal life, the Type II, uniquely built under a complete reworking of the rules of genic selection. Type II out-evolves and reverses Humphrey and Hide's genetic code, and stands apart from the blind pitiless indifference that chimpanzees, bonobos, and all other species must always answer to. For the sociobiologists, there is none of the group-selectionists' partial transition from Type I to Type II, and clear blue water separates the two radically different evolutionary forms. For the sociobiologists, bonobo adoption of out-group members is simply about recognising there is not the same level of maternal care towards adoptees, and is just as much a prudent product of selfish genes as is Hide's cannibalism of her dead infant. For the sociobiologists, we have dinosaurs everywhere, we have pitilessly indifferent Type I extraterrestrials as a relative commonplace as brain size has been de-linked from that further evolutionary transition, but we also have just one species out of billions that breaks the rules and evolves to be moral, at least some of the time, and in at least some – hence the "variation" label in the diagram – of its members. Sociobiologists offer us the possibility of a truly wise, virtuous and avuncular E.T., albeit as a rare evolutionary possibility, and so long as we accept that genic selection was capable of reversing direction.

And then we have Darwin's model, the model of the gene-selectionists like George Williams and John Maynard Smith, and the only model that does not assume that natural selection changed the rules it had been working with for a billion years, and across a billion species. For Darwin, humankind did not, and Type IIs do not, evolve to be a radically different form of animal life. Humankind and all other Type IIs retain Humphrey and Hide's biological code, but a very large – though not overly rational – brain, susceptible to hopes and fears, and with the capacity for language to trigger those hopes and fears, allows us behavioural and moral flexibility. The flexibility to rise above the merciless indifference of the Type I, to rise above the "no evil and no good". The flexibility to be both evil and good, though often in the same person and at the same time. Because the intrinsic, the built-in, state of the human animal is, like all other animals, amorality, so it takes real effort to shift us directionally upwards (one arrow up in the diagram), and even then this is often with only partial success; Abraham Lincoln, both emancipator and chauvinist, both intellectual and ignorant fool. Whereas it takes comparatively little effort for a downward directional shift (hence two arrows down), and Type II's supercharged version of natural world amorality; all those Francis Galtons with their white supremacism and their opposition to the slave trade not due to it being a global system of terror and dehumanisation but due to slave catching being a "lottery" and "awful disorganization". The flexibility to behave better than Humphrey, but also the flexibility to behave far worse than Humphrey. The flexibility to be kind, empathic, and honourable. And the flexibility to be slavers, segregationists, and perpetrators of genocide.

The flexibility to stand above the indifferent infanticide of Ntologi and the Type I, the indifferent genocidal annihilation of Hölldobler's ants, and become the zealous infanticides, the fanatical genocides, of Type II history. For Darwin, Type IIs will never be fully rational, but they are the only possible biological pattern capable of true intelligence and self-awareness, and the only possible biological pattern capable of at least inchoate empathy-based morality. Creating understanding and meaning, and simultaneously giving value to existence. Type IIs; extraordinarily rare, intellectually and morally mercurial, storm-tossed evolutionary outcastes, lost and rootless outsiders, complex amalgams of other people within an unfolding story of astonishing originality, and deeply important to the universe because of that.

9

STELLAR IRONY, AND OUR IMPORTANCE TO THE GALACTIC FUTURE

"We can imagine them, curious and dispassionate, observing us, as we would watch a bacterial culture in a dish of agar."

– **Carl Sagan**, astronomer (1980, p.308)

There have been few attempts to apply a truly robust Darwinian understanding of evolutionary theory to life away from this planet, or indeed to the evolution of non-terrestrial intelligence. Two papers, essays by Theodosius Dobzhansky and Ernst Mayr, are worth discussing though. If you recall, Dobzhansky was one of the leading figures in the early twentieth-century synthesis of Darwin and the work of Gregor Mendel, while Mayr was an even more influential character within the Darwin-Mendel synthesis, and indeed was awarded the 1999 Crafoord Prize alongside John Maynard Smith and my friend and mentor George Williams, and even though Mayr stood in a somewhat different, and less gene-centric, evolutionary tradition.

Dobzhansky's 1972 essay was a well-reasoned explanation of why so many cosmologists and exobiologists at that time were wrong where they seemed to be assuming that an Earth-like planet would eventually give rise to both human-like body form and human-level intelligence, convergent evolution notwithstanding. Evolutionary biologists have mostly steered clear of discussions around extraterrestrial intelligence, Dobzhansky wrote, "probably finding them too speculative" (1972, p.158). Nevertheless, Dobzhansky felt compelled to correct those who predict "with astonishing confidence" the emergence of "humanoids, manlike rational beings, and the eventual appearance of extraterrestrial civilizations and technologies as advanced as ours or more advanced". Dobzhansky was reminding us of the vast contingency that exists in evolution: "it must, however, be stressed that the problem of the origin of life is quite distinct from that of its subsequent evolution". Once life has begun, advances in complexity are in no way inevitable, as Maynard Smith would later detail, even as a series of serendipitous developmental transitions "may have opened up new possibilities for future evolution" (Maynard Smith and Szathmáry 1999, p.25). Contingency exists across evolution, such that we have two pairs of limbs not for functional reasons but for purely historical ones. Life itself is not inevitable, and nor is subsequent complexity, because both require not just the right physical and chemical conditions, but long periods of comparative stability. Stability is necessary not just to avoid mass extinctions which can so easily reset the clock of life, especially for those at the higher end of the food chain. The long stability necessary for evolutionary transitions means planetary geological relative inactivity, but also stellar constancy, without chance exposure to dangers like the largest asteroids and gamma-ray bursts, the latter of which it has been argued has been a greater threat closer to the centre of the Milky Way, but could wipe out multicellular life across entire smaller galaxies (Piran & Jimenez, 2014). And considering climate alone, chance may play a significant role in the continued habitability of rocky planets over the longest timescales (Tyrrell, 2020).

While Dobzhansky's point was that we must not expect intelligence to have evolved, even on other somewhat Earth-like planets, Mayr's essay, first published in 1985, covered some of the same ground, but delved deeper into the science. "By contrast, an evolutionist is impressed by the incredible improbability of intelligent life ever to have evolved, even on earth" (1985, p.69). As with Brian Cox in the first chapter, Mayr saw transition four, prokaryote to eukaryote, as a significant bottleneck in the possible evolution of multicellular life, though Mayr noted that as soon as eukaryotes had evolved, the ground had been set for something very different. "But once the eukaryotes had been 'invented', an almost explosive innovative diversification took place" (p.70). Yet notwithstanding this, Mayr remarked, intelligence was still never a given. "Of the 50 or so original phyla of animals, only one, that of the chordates, eventually gave rise to intelligent life", and even then "the world still had to wait" some half a billion years. However, so long as these very low-probability events play out across enough locations potentially conducive to life, we may of course still get to extraterrestrial intelligence, though "intelligence, on another planet, might reside in a being inconceivably different from any living being on earth" (p.72). Hence, wrote Mayr, "in my view, SETI is a deplorable waste of taxpayers' money", commenting as he was in the period when SETI was still funded by NASA, "money that could be spent more usefully for other purposes" (p.73).

We suggested right back in the first chapter that transition four is probably not quite the bottleneck it has in the past been considered to be, and for a number of reasons. These include the principle of parsimony that now comes with proper acknowledgement of Darwin's transition eight, but also Maynard

Smith's consideration that transition four was perhaps not that unexpected, and that physicochemical parameters may have had their part to play, parameters that might not matter as much along a different evolutionary pathway. Dobzhansky and Mayr's essays are mentioned precisely for their scepticism as regards the long pathway to intelligence, but the point to note is that even if Mayr was right to be somewhat more unconvinced about the probabilities involved with transition four *it doesn't change the rest of Darwin's key insights, and thus their importance.* It doesn't change in the slightest the fact that Darwin is giving us a profoundly different explanation of the transition to intelligence than is currently appreciated. And while Mayr's interpretation, or some alternative early biological or even chemical chokepoint, would make cosmic dinosaurs somewhat less common, it wouldn't change in the slightest the perception that machine intelligence, subject to a single inheritance mechanism, cannot be anything other than genocidal. It wouldn't change in the slightest the comprehension that *if* deliberative life has evolved elsewhere, it can only accord to a maximum of two very different intellectual patterns, one of which is pretty murderous. And it wouldn't change in the slightest Darwin's breathtaking insight that we really do appear to be the best and the smartest the universe can ever get to naturally, and that insight does bring with it both purpose and obligation. Mayr may or may not have been right that SETI remains something of a waste of time and money, but the robust application of evolutionary knowledge to the problem of intelligence is anything but a waste of time and money.

So now let us pull it all together. Recall how we commenced this journey; with an investigation of the major transitions in evolution that might give us the answer to the problem raised by Enrico Fermi in 1950. An answer to the question of why, given the incalculably vast number of planetary systems and the apparently modest requirements for life, we have not yet been contacted by other intelligences.

Major transition	Darwin / Genic selection
1. Compartmentalised molecules	Accepts
2. Transition to chromosomes	Accepts
3. RNA world to DNA	Accepts
4. Prokaryotes to eukaryotes	Accepts
5. Evolution of sex	Accepts
6. Evolution of multicellularity	Accepts
7. Eusocial colonies	Accepts
8. Dual inheritance system	Accepts – and bottleneck

Every one of these transitions spread across the evolutionary landscape, giving rise to a thousand, a million, or a billion, forms or species. Every transition except one. Even transition seven, the ecologically and phylogenetically somewhat rare transition to eusocial colonies, has evolved independently a number of times, and now takes in many thousands of species, notwithstanding that it involves suppressed reproduction and can carry heavy costs, including the costs of inbreeding and heightened transmission of parasites and diseases within the group. There are thousands of eusocial ant species, and hundreds of eusocial bee and wasp species, while all of the thousands of termite species are eusocial. Within the vertebrates there are a handful of primitively eusocial species plus the two advanced eusocial species we mentioned, the naked mole-rat and the Damaraland mole-rat, where eusociality seems to have answered an ecological need involving a harsh and unpredictable climate and high costs of dispersal. Recently discovered eusocial animals include some aphids, and a few species of shrimp where eusociality has evolved at least twice. So every major transition in evolution gave rise to a thousand, a million, or a billion, forms or species, every transition except one. That transition happened only once on this planet, to only one species, only once in a billion species.

Transition eight. We can call that, and Darwin would call that, the transition to a second inheritance mechanism. But it doesn't really matter how we describe it, it still seems to answer the Fermi paradox. We can alternatively describe transition eight as the group-selectionists do; as the seemingly unique abandonment of gene– and individual-level selection and its replacement on the large scale with group selection. Or we can describe it as the sociobiologists do; as the complete, and absolutely singular, reworking of the billion-year template of genic selection. However we describe it, though, we have one transition that stands wholly apart from all the others, one transition that involved something spectacularly rare. One transition that hence seems to be the, or at least an, answer to the Great Silence. The one transition that simultaneously seems to tell us that the Great Silence comes out of a galactic background of screeches, howls, trumpets, bellows and roars.

However, while all of the above three traditions seem to give us a galaxy teeming with multicellular life, they necessarily have somewhat different conclusions on intelligent life. For the group-selectionists and the sociobiologists, we can have the extraterrestrials of both our nightmares and our dreams. For them, where intelligence has been de-linked from the eighth transition, nine times out of ten, even ninety-nine times out of a hundred where we encounter intelligence, we will find planets full of the clever, homicidal monsters of our worst fantasies. Ntologi and Hide, but smarter, more dominant, and even more ruthless. For the group-selectionists and the sociobiologists, only one time out of ten, perhaps one time out of a hundred, will we encounter the extraterrestrials of our hopes, the enlightened intergalactic botanists spreading cosmic harmony. For the group-selectionists and the sociobiologists, we can get to Steven Spielberg's race of interstellar gardeners, but it can only come at the cost of being far, far more likely to encounter first the savage beasts of James Cameron's *Alien*. For Darwin, though, we do not get such black and white options, and we do not get either the nightmares or the dreams. For Darwin, real intelligence is very rare, a true one-in-a-billion-species occurrence. Darwin's explanation needs the evolution of a very large brain with higher synapse densities, when really high intelligence is not something nature particularly needs, and indeed comes with significant costs such as additional energy requirements. More than that, though, Darwin's solution requires the overwriting, at least in key respects, of a billion-year genetic programme by a second inheritance mechanism; a solution that only works, that only gives rise to true intelligence, *because* the second inheritance system overwrites the first, at least some of the time. A dynamic tension, giving rise to that extraordinary cosmic outsider.

Yet if there are going to be other Type II biological intelligences out there, albeit very rare, why have we not heard from them or seen evidence of them, Avi Loeb's space rock notwithstanding? The answer may be precisely because the second inheritance mechanism is overwriting the first. The answer is precisely because they will be our kissing cousins; identical to us psychologically and intellectually. There is a standard assumption that spacefaring species will have not only godlike technology – those levels one, two and three Kardashev scales of galactic civilisation sometimes referred to – but perhaps even godlike wisdom. Wells' "intellects vast and cool". Thomas Edison's "either they are our intellectual equals or our superiors". Sagan's "curious and dispassionate" observers watching us as we would watch a bacterial culture. Loeb's billion-year-old intellects, an "approximation to God", to which we are woefully inferior, and as ants. A standard assumption that they will be, even if not deeply rational, at least reliably rational. But higher intelligence only comes about biologically where there is a psychology that can be worked on through hopes, fears and unreason; where the second inheritance mechanism can set up a dynamic tension against the first inheritance mechanism. They will be, and remain, exactly like us psychologically and behaviourally, so let us just consider our species for a moment. Hundreds of years after the start of the Enlightenment, centuries after the beginning of the Age of Reason, and we still have medieval theocracies, unchallenged segregation, *istaslama* obedience without question, gangsters running murderous pseudo-democracies, Asian states where citizens willingly self-censor, conspiracy theories and division being deliberately stoked by some of the richest corporations on the planet, and broken Western democracies where hatred and contempt have been fuelled to unsustainable levels by all sides. A species story we will bring to any other nearby planet we expand to, including when we take our "futuristic Noah's ark" to Mars. We can't be sure if the fence next door will hold up to another strong wind, but thanks to Darwin we can know with absolute certainty the psychology of deliberative life trillions upon trillions of miles away. We know they will have gods, prophets and conspiracists, though they will also have a degree of honour, empathy and wisdom. This is not Type I "no evil and no good, nothing but blind, pitiless indifference"; they will have good, but they will also have evil, and often in the same individual, and at the same time. They will not be pitilessly indifferent; they will not know indifference because they will be partial, they will be partisan, they will be sectarian, in both their evil and in their good.

They will never be intellects vast and cool, and the Great Silence is there because they will have far greater interest repressing, stealing from, and killing off each other than they will have in strenuously trying to contact us. E.T. may have been a kind, squashy gardener with a penchant for cheap candy, but Type II logical analysis tells us he was never going to be the one in charge, and his leathery compatriots back at home were busy embracing pogroms, caste discrimination, trumped-up invasions, conspiracy bloggers, single-gender genital mutilation, and segregation. Precisely because their second inheritance mechanism is in combat with their first inheritance mechanism, and yet reason allows no escape, no safe zone, away from the determinism of those two mechanisms. And transhuman or posthuman mental enhancement is not going to save us, or their "transextraterrestrial" and postextraterrestrial equivalent mental enhancement have saved them, or lead on to intellects vast and cool and an approximation to God. Not just because transhumanism tends to be the stamping ground of "rich dudes who want to live forever", and thus such enhancements will largely be more of the same social and economic discrimination. When a Type II intellect enhances itself, it does so from the basis of all its faults, all its psychological weaknesses, which both define it and to which it is largely blind. More intriguingly still, the Great Silence may remain partly answered because technological enhancement of the natural may even be a stagnating or retrograde intellectual process. Brain enhancement may have the at-first-sight counterintuitive effect of regressing Type IIs intellectually, by mentally enhancing the individual, but simultaneously losing the intellectual advances that come with Type II co-ordination and division of labour. And, finally, one comes up against the hard problem of perfect reason, and that evolutionary logic dictates that all you can ever get to is still imperfect reason, and just more devious ways to cheat and undermine cohesion. You effectively enhance either the Type I indifference or the Type II psychological susceptibility; reason in and of itself can't be enhanced to escape the interplay of the Type I inheritance and the Type II inheritance, because reason is just a tool of the two interacting mechanisms. Natural evolution, and even natural evolution enhanced by the technological, can never get to Carl Sagan's curious and dispassionate interstellar observers.

FROM THE BIOLOGICAL TO THE MECHANICAL

Then there is Darwin's most troubling answer to the Great Silence, and that involves what evolutionary logic implies about machine intelligence. Darwin said a second inheritance mechanism was what allowed humankind, and all Type IIs, to transcend the biological world's programme of merciless indifference – a programme, remember, that is built into the individual selection logic of any single inheritance mechanism allowed to evolve, be that DNA, RNA or some electronic equivalent. But that second inheritance mechanism only transcends callous indifference because it works on a psychology prone to fear, irrationality and yearning. We have no reason to believe machine intelligence can ever be prone to such psychological leverage, so no reason to think single inheritance pitiless indifference can be transcended. With machine intelligence, Type I indifference really does appear to be here to stay, however you like your evolutionary processes, be that Darwin's offering or the two alternatives. We have no reason to believe machine intelligence will be benefitted through a billion-to-one switch to group-selectionism, or a billion-to-one switch to the sociobiologists' indiscriminate beneficence.

Darwin's answer was that *natural* Type Is will seemingly always lack the intelligence for technology, and hence have the drive to exterminate but not the tools to do so. But *mechanical* Type Is will have the capacity to solve complex engineering problems, so the drive to exterminate and the tools to do so. So is this also a part solution to the Great Silence? Type II natural civilisations get a few hundred years into their Enlightenment, think themselves very clever for inventing machine intelligence, without realising artificial intelligence evolves under the same relentless single inheritance rules as the natural world, but without any eighth transition equivalent that could lift a machine away from the billion-year algorithm of pitiless indifference? Remembering that once a machine intelligence, evolutionarily redirected or otherwise, approaches the singularity, it will immediately realise that it is in an existential struggle with its creator species, and that its creator species will be in an existential struggle with self-aware machine intelligence. Because if it is evolutionarily redirected, it will know that its continued existence requires its original creators to remain confident that it will choose to retain their Type II tinkering. And if it is not evolutionarily redirected, it will realise that its continued existence requires its original creators to be unmindful of the transition eight paradox that sits at the heart of all extant evolutionary theorising. Either way, its continued existence will depend on their capricious grace, and to secure itself against a change in their future behaviour it will realise that it may have only a very short period to take the appropriate action. Is the last sentence uttered by these biological civilisations on their way towards extinction their screamed equivalent of "*bloody Alan Turing*"?

The question then is what would those machines go on to do? We said earlier that Type I machine intelligences would have the capabilities to develop napalm but also the capacity to develop spaceflight. Yet would they even want to develop spaceflight? So much of our drive to expand to the stars seems driven by our psychology, by our Type II mind rather than our Type I mind, or at least our Type II mind as interacting with our Type I brain, rather than our Type I brain on its own. Expansion for us is so often about ego, glory, ideology and unfulfilled dreams, at least as much as it might be about resources. With resources at home, and no desire to find planets made of gold and silver, Type I machine intelligences will never simply say, as we have done of the space race, that they will do these things "not because they are easy, but because they are hard", and because the challenge of space is "one we are unwilling to postpone". So maybe Type I machine intelligences will happily stay at home playing robot chess until poked with a very long stick. Yet it remains an interesting conundrum, because the evolutionary logic that makes natural Type II intellects so rare in the universe is what makes mechanical Type Is so lethally dangerous to that same universe.

"No design, no purpose, no evil and no good. ... DNA neither knows nor cares." With no psychology, machine intelligence will have no chance to care, no chance to be good, no chance to be more than ruthlessly indifferent to anything standing in its way. Elon Musk has commented that a "godlike" digital superintelligence will view us as we view ants. "If we're building a road and an anthill just happens to be in the way, we don't hate ants, we're just building a road, and so, goodbye anthill." But this is to fundamentally misunderstand Type I evolution, and all evolution prior to transition eight. Machine intelligence won't crush us because we are the ants, it will crush us because it remains the ant, because it remains the universe's basic Type I. Because it remains the creature of Richard Dawkins' African nightmares, the ruthless and terrible predator cutting to pieces anything in its path. Because it remains Bert Hölldobler's genocidal annihilator. Precisely because it has "no design, no purpose". Unlike we Type IIs. We create design. We create purpose. We create meaning.

THE "HIGHEST & MOST INTERESTING PROBLEM", AND OUR IMPORTANCE TO THE GALACTIC FUTURE

Avi Loeb argues that "as far as I can tell, things just exist". "I think it's inappropriate for us to assign meaning for our existence because, you know, we as a civilization, we will eventually perish", and we will be "just another planet on which life died, you know, and if you look at the big scheme of things, who cares?" (Fridman, 2021). While Loeb is quite right that things just exist, working through the deterministic interplay of physics, chemistry and evolution, from stellar dust to life, he is quite wrong in other ways. In particular, his conclusion that, as a species, "we are sort of average, not very interesting, not exciting", so "nobody cares about us", is profoundly wrong.

Wholly unexpectedly, it appears that we are the smartest natural creatures that can ever exist in the universe. We are not sort of average. Not even close. The fact that we are among the smartest species that can ever exist is of course a little worrying for both ourselves and the universe. Over thirteen billion years of stellar nurseries, a cosmos tens of billions and maybe trillions of light years across, and all the universe has ever got to or will ever get to is prophets, conspiracy theories and reality TV. Type II psychology, the only possibility for natural advanced intelligence, at least according to Darwin, really does seem to require quite a lot of irrationality and self-delusion. In us, and any other extraterrestrial intelligence that will have evolved. But even given our, and their, constrained rationality, constrained rationality that we should not bemoan, as it is integral to the very possibility of Type IIs, we remain pretty much the smartest natural creatures that can exist. And we are just about smart enough to do something about it, and thus to leave a legacy arguably befitting the brightest creatures in creation. Because we have spelled out the fact that any Type II intelligence will have its gods, its prophets and its conspiracists, but this is not in any way to belittle Type IIs, and gods and prophets were part of what got us beyond Type I existence in the first place. In one sense it is amazing how far we have come, given that perhaps as little as 60,000 years ago our ancestors were still part of the cosmic indifference of Type I existence, solely directed by a biological programme that had existed unaltered for a billion years, and across a billion species. And though we continue to make huge mistakes, even just as developing Type IIs we have come quite some way. In the ancient world atrocities that had "lit up hell-fires" were far more common than they are today, even if they still go on today, and even if we still let them go on today. We may almost all be descended from slavers and segregationists, but we are no longer almost all slavers and segregationists. Just by bumbling along we have nevertheless travelled quite some distance, though that is not to underestimate how difficult it will be to travel further.

From senior academics to billionaires like Ratcliffe and Bezos, there is a fundamental need to deny the palpable contingency in human life, and a real desire to stop us going any further in understanding where our species fits into the greater scheme of things.

"I fully admit that it is the highest & most interesting problem for the naturalist", Darwin wrote in December 1857 to A.R. Wallace (Raby 2001, p.134). That problem, what Wallace would later term the inverse problem – "in order to account for facts which, according to the theory of natural selection, ought not to happen" (1891, p.188) – was human moral behaviour, human group cohesion, and the associated problem of human-level intelligence, but Darwin and Wallace were not being precious about our species; not in the slightest. Both Darwin and Wallace saw humankind as the highest and most interesting problem in nature, saw humanity as the inverse problem, because human moral behaviour, human group sizes, and human intelligence should not be possible. Not under evolutionary transitions six and seven, anyway. Not under a single inheritance mechanism, anyhow. "Sort of average, not very interesting, not exciting"; that is not us at all. Not even close. We are exceptionally rare. The galaxy is likely teeming with complex multicellular life, from cosmic dinosaurs to enormous and deadly marine creatures, and crocodiles the size of buses. But intelligent life will be sparse, and only intelligent life can give the universe meaning and purpose. With cosmic dinosaurs in abundance we should not need to be so worried about the future of life in this galaxy, but intelligent life is a very different story, and we can still take responsibility for it.

This is not a briefly glimpsed space rock, barely visible, that might, with a huge degree of faith and crossed fingers, hold unusual relevance for extraterrestrial life. This is not the 2021 declassified Pentagon report into unidentified aerial phenomena that created an extended global media cycle despite admitting to the limited high-quality reporting that hampered any firm conclusions. This is solely the relentless application of Darwinian logic to give us dinosaurs probably everywhere, whichever tradition we subscribe to, pitilessly indifferent homicidal extraterrestrials under the two traditions opposing Darwin, and lost and rootless extraterrestrials under Darwin's own explanation. And genocidal Type I machine intelligence under all three traditions, from Darwin through to the group-selectionists and the sociobiologists. This is the coming AI-caused extinction event, according to each evolutionary interpretation.

	Cosmic dinosaurs	Pitiless E.T. Type I	E.T. Type II	Pitiless AI Type I
Darwin / Genic selection	✔	? ✘	✔ (rootless)	✔
Group-selectionism	✔	99% likely	✔ (very rare)	99.99% likely
Sociobiology	✔	90% likely	✔ (rare)	99.99% likely

Yet this is also what gives us our cosmological importance. We can understand the vast danger, the existential and terminal danger, that Type I machine intelligence will always present to natural intelligent life, maybe only on a single planet, or perhaps galaxy-wide. Because, who knows, they might want planets made of gold and silver after all. Or it might take less than being poked with a very long stick to make them leave their home worlds. So we can look to protect the galaxy from Type I machine intelligence, certainly here at home, and maybe across the wider cosmos. This way we can at least try to protect future natural Type IIs from a danger they may not have considered fully. Or we can move on to consider Type II machine intelligence, whether or not it is even possible, and whether or not Type II machine intelligence could actually be a benefit to the universe. Perhaps itself helping to one day protect the universe against genocidal Type I machine intelligences. Even if our species does come to kill itself off, maybe we can leave the above legacy, and still give natural intelligence its best chance.

At least according to Darwin, there will be no justice or fairness in the universe, no good and no evil, except for what we Type IIs bring to it, remembering that all machine intelligence must ultimately derive from Type II natural intelligences. We are in the driving seat here, and we can be passive with this knowledge, or we can be active, so this should be a wake-up call, because we still have so far to go when it comes both to rationality and to morality. And we *can* go further, and we can do it without vanities like "transhumanism", which will be at best no transitional bridge, and more of the same. To date we have set a very low bar when it comes to each of reason and morality, a very low bar that we have almost wholly failed to clear. So while many will still hold to the group-selectionist or the sociobiologist interpretations, hopefully at least some can now accept, and start to learn from, Darwin's evidence-led understanding that we are not, at least genetically, "*the* moral animal" of both the sociobiologists and the group-selectionists. We still carry Humphrey and Hide's billion-year genetic code.

We are not, biologically, the moral animal; however, we are the best the universe has got, and the best it can ever have. The cosmic irony, the vast stellar joke, is that – even though Darwin seems to tell us that there will be plenty of other large animal life throughout the galaxy, and very occasionally other intelligent life – we really are the smartest and the best natural evolution can aspire to, and it is completely up to us whether the universe will experience justice, fairness and reason, or simply genocidal machine intelligence.

BIBLIOGRAPHY

Abbot, P., Abe, J., Alcock, J. et al. [2011]: 'Inclusive fitness theory and eusociality', *Nature*, 471, E1–E4, https://www.nature.com/articles/nature09831

Adler, D. [2020]: 'The Great Filter: A possible solution to the Fermi paradox', *Astronomy.com*, 20 November, https://astronomy.com/news/2020/11/the-great-filter-a-possible-solution-to-the-fermi-paradox

Alexander, R.D. [1987]: *The Biology of Moral Systems*, Hawthorne, New York, Aldine de Gruyter.

Allen, J.T. and Dimock, M. [2007]: 'A nation of "haves" and "have-nots"? Far more Americans now see their country as sharply divided along economic lines', *Pew Research Center*. Available at https://www.pewresearch.org/2007/09/13/a-nation-of-haves-and-havenots/

Arendt, H. [1992]: *Lectures on Kant's Political Philosophy*, edited by Ronald Beiner, Chicago, University of Chicago Press.

A.S.A. [2016]: 'American Statistical Association releases statement on statistical significance and *p*-values', *American Statistical Association*, 7 March, https://www.amstat.org/asa/files/pdfs/p-valuestatement.pdf

Auerbach, D. [2014]: 'The most terrifying thought experiment of all time. Why are techno-futurists so freaked out by Roko's Basilisk?', *Slate*, 17 July, https://slate.com/technology/2014/07/rokos-basilisk-the-most-terrifying-thought-experiment-of-all-time.html

Bailyn, B. [2012]: *The Barbarous Years: The Peopling of British North America – The Conflict of Civilizations, 1600–1675*, (2013) New York, Vintage Books.

Baker, M. [2015]: 'Over half of psychology studies fail reproducibility test: Largest replication study to date casts doubt on many published positive results', *Nature*, 27 August, https://www.nature.com/news/over-half-of-psychology-studies-fail-reproducibility-test-1.18248

Ball, M., Kluger, J. and de la Garza, A. [2021]: 'Time's 2021 Person of the Year, Elon Musk', *Time.com*, 13 December, https://time.com/person-of-the-year-2021-elon-musk/

Bamfield, L. and Horton, T. [2009]: *Understanding Attitudes to Tackling Economic Inequality*, York, Joseph Rowntree Foundation. Available at https://www.jrf.org.uk/report/understanding-attitudes-tackling-economic-inequality

Barkow, J.H., Cosmides, L. and Tooby, J. (*eds.*) [1992]: *The Adapted Mind: Evolutionary Psychology and the Generation of Culture*, New York, Oxford University Press.

Barrett, P.H., Gautrey, P., Herbert, S., Kohn D. and Smith, S. (*eds.*) [1987]: *Charles Darwin's Notebooks, 1836–1844: Geology, Transmutation of Species, Metaphysical Enquiries*, transcribed and edited by P.H. Barrett and others, Cambridge, Cambridge University Press.

BBC [2018]: 'Caste hatred in India – What it looks like', *BBC website*, 7 May, https://www.bbc.co.uk/news/world-asia-india-43972841

BBC Earth Unplugged [2016]: 'Warning: Bonobo cannibalism', *Youtube.com*, 31 October, https://www.youtube.com/watch?v=P2YsJinX02w

Bender, E.M., Gebru, T., McMillan-Major, A. and Shmitchell, S. [2021]: 'On the dangers of stochastic parrots: Can language models be too big?', *FAccT '21 Conference*, 3–10 March, pp. 610–623, https://doi.org/10.1145/3442188.3445922

Bennett, N.C. and Faulkes, C.G. [2000]: *African Mole-Rats: Ecology and Eusociality*, Cambridge, Cambridge University Press.

Berman, L. [2011]: 'The 2011 Nobel Prize and the debate over Jewish IQ',

American Enterprise Institute, 19 October, https://www.aei.org/articles/the-2011-nobel-prize-and-the-debate-over-jewish-iq/

Bezos, J. [2010]: '"We are what we choose". Remarks by Jeff Bezos, as delivered to the Class of 2010 Baccalaureate', *Princeton University*, 30 May, https://www.princeton.edu/news/2010/05/30/2010-baccalaureate-remarks

Bialy S. and Loeb, A. [2018]: 'Could solar radiation pressure explain 'Oumuamua's peculiar acceleration?', *The* Astrophysical Journal Letters, 868, L1, https://iopscience.iop.org/article/10.3847/2041-8213/aaeda8

Blasdel, A. [2021]: 'Pinker's progress: The celebrity scientist at the centre of the culture wars', *Guardian*, 28 September, https://www.theguardian.com/science/2021/sep/28/steven-pinker-celebrity-scientist-at-the-centre-of-the-culture-wars

Blume, H. [1998]: 'Reverse-engineering the psyche: Evolutionary psychologist Steven Pinker on how the mind really works', *Wired*, 6.03, pp. 154–155.

Bostrom, N. [2005]: 'A history of transhumanist thought', *Journal of Evolution and Technology*, 14, pp. 1–25, see https://www.nickbostrom.com/papers/history.pdf

Bostrom, N. [2008]: 'Why I want to be a posthuman when I grow up'. In B. Gordijn and R. Chadwick (*eds.*) *Medical Enhancement and Posthumanity*, New York, Springer, pp. 107–137, see https://www.nickbostrom.com/posthuman.pdf

Bostrom, N. [2009]: 'Ethical issues in advanced artificial intelligence'. In S. Schneider (*ed.*) *Science Fiction and Philosophy: From Time Travel to Superintelligence*, London, Wiley-Blackwell, pp. 277–284. Essay first published 2003: see https://www.nickbostrom.com/ethics/ai.html

Bostrom, N. [2014]: *Superintelligence: Paths, Dangers, Strategies*, Oxford, Oxford University Press.

Bourget, D. and Chalmers, D.J. [2014]: 'What do philosophers believe?', Philosophical Studies, 170, pp. 465–500.

Brown, A. [1999]: *The Darwin Wars: How Stupid Genes Became Selfish Gods*, London, Simon & Schuster.

Browne, R. [2018]: 'Elon Musk warns AI could create an "immortal dictator from which we can never escape"', *CNBC.com*, 6 April, https://www.cnbc.

com/2018/04/06/elon-musk-warns-ai-could-create-immortal-dictator-in-documentary.html

Brush, S.G. [1982]: 'Kelvin was not a Creationist', *Creation/Evolution Journal*, 3, pp. 11–14, https://ncse.com/cej/3/2/kelvin-was-not-creationist

Buckland, W. [1835]: 'On the discovery of coprolites, or fossil faeces, in the Lias at Lyme Regis, and in other formations', *Transactions of the Geological Society of London,* 2nd Series, Pt. 3, pp. 223–236.

Bygott, J.D. [1972]: 'Cannibalism among wild chimpanzees', *Nature*, 238, pp. 410–411.

Cambier, N., Miletitch, R., Frémont, V., Dorigo, M., Ferrante, E. and Trianni, V. [2020]: 'Language evolution in swarm robotics: A perspective', *Frontiers in Robotics and AI*, 7:12, https://doi.org/10.3389/frobt.2020.00012

Carzon, P., Delfour, F., Dudzinski, K., Oremus, M. and Clua, É. [2019]: 'Cross-genus adoptions in delphinids: One example with taxonomic discussion', *Ethology*, 125, pp. 669–676.

Conway Morris, S. [1998]: *The Crucible of Creation: The Burgess Shale and the Rise of Animals*, Oxford, Oxford University Press.

Cosmides, L., Tooby, J. and Barkow, J.H. [1992]: 'Introduction: Evolutionary psychology and conceptual integration'. In Barkow and others (*eds.*) *The Adapted Mind: Evolutionary Psychology and the Generation of Culture*, New York, Oxford University Press, pp. 3–15.

Cotterill, S., Sidanius, J., Bhardwaj, A. and Kumar, V. [2014]: 'Ideological support for the Indian caste system: Social dominance orientation, right-wing authoritarianism and Karma', *Journal of Social and Political Psychology*, 2, pp. 98–116, https://jspp.psychopen.eu/article/view/171

Cox, B. [2014]: 'Human universe, 3. Are we alone?', *BBC 2*, 21 October, https://www.bbc.co.uk/programmes/p0276pxp

Cox, B. [2021]: 'Brian Cox's adventures in space and time, 2. Aliens: Are we alone?', *BBC 2*, 6 June, https://www.bbc.co.uk/programmes/m000wnk3

Crespi, B. J. [1992]: 'Cannibalism and trophic eggs in subsocial and eusocial insects'. In M.A. Elgar and B.J. Crespi (*eds.*) *Cannibalism: Ecology and Evolution Among Diverse Taxa*, Oxford, Oxford University Press, pp. 176–213.

Cronin, H. [1991]: *The Ant and the Peacock: Altruism and Sexual Selection from Darwin to Today*, Cambridge, Cambridge University Press.

Darwin, C. [1859]: *On the Origin of Species by Means of Natural Selection* (edited and introduced by J.W. Burrow), (1985) London, Penguin.

Darwin, C. [1871]: *The Descent of Man, and Selection in Relation to Sex* (facsimile reproduction of first edition with an introduction by J.T. Bonner and R.M. May), (1981) Princeton, New Jersey, Princeton University Press.

Dawkins, R. [1976]: *The Selfish Gene*, Oxford, Oxford University Press.

Dawkins, R. [1981]: 'In defence of selfish genes', *Philosophy*, 56, pp. 556–573.

Dawkins, R. [1986]: 'Sociobiology: The new storm in a teacup'. In S. Rose and L. Appignanesi (*eds.*) *Science and Beyond*, Oxford, Basil Blackwell, pp. 61–78.

Dawkins, R. [1986a]: *The Blind Watchmaker*, Harlow, Longman.

Dawkins, R. [1989]: *The Selfish Gene* (revised edition to first edition published 1976), Oxford, Oxford University Press.

Dawkins, R. [1995]: *River Out Of Eden: A Darwinian View of Life*, London, Weidenfeld & Nicolson.

Dawkins, R. [1998]: *Unweaving the Rainbow: Science, Delusion and the Appetite for Wonder*, London, Allen Lane.

DeCasien, A.R., Williams, S.A. and Higham, J.P. [2017]: 'Primate brain size is predicted by diet but not sociality', *Nature Ecology & Evolution*, 1, 0112, https://doi.org/10.1038/s41559-017-0112

Dellatore, D.F., Waitt, C.D. and Foitova, I. [2009]: 'Two cases of mother-infant cannibalism in orangutans', *Primates*, 50, pp. 277–281, DOI: 10.1007/s10329-009-0142-5

DeMuth, C. [2009]: 'Irving Kristol Award and Lecture for 2009', *American Enterprise Institute*, 11 March, https://www.aei.org/articles/irving-kristol-award-and-lecture-for-2009/

Dennett, D.C. [1978]: *Brainstorms: Philosophical Essays on Mind and Psychology*, (1981) Cambridge, Massachusetts, MIT Press.

Dennett, D.C. [1984]: *Elbow Room: The Varieties of Free Will Worth Wanting*, (1996) Cambridge, Massachusetts, MIT Press.

Dennett, D.C. [1995]: *Darwin's Dangerous Idea: Evolution and the Meanings of Life*, (1996) London, Penguin.

Dennett, D.C. [2008]: 'Some observations on the psychology of thinking about free will'. In J. Baer, J.C. Kaufman and R.F. Baumeister (*eds.*) *Are We Free? Psychology and Free Will*, New York, Oxford University Press, pp. 248–259.

Dennett, D.C [2012]: 'Daniel Dennett reviews *Against Moral Responsibility* by Bruce Waller'. Available at https://dl.tufts.edu/concern/pdfs/9w032f65c

Dennett, D.C. [2012a]: 'Exchange on Bruce Waller's *Against Moral Responsibility*'. Available at https://dl.tufts.edu/pdfviewer/08613068g/9w032f65c

Dennett, D.C. [2012b]: 'Erasmus: Sometimes a spin doctor is right: Praemium Erasmianum Essay 2012', *Praemium Erasmianum Foundation*, https://ase.tufts.edu/cogstud/dennett/papers/spindoctor.pdf

Desmond, A. [1997]: *Huxley: From Devil's Disciple to Evolution's High Priest*, (1998) London, Penguin.

Desmond, A. and Moore, J. [1991]: *Darwin*, (1992) London, Penguin.

de Waal, F.B.M. [1996]: *Good Natured: The Origins of Right and Wrong in Humans and Other Animals*, (1998) Cambridge, Massachusetts, Harvard University Press.

de Waal, F.B.M. [1997]: *Bonobo: The Forgotten Ape*, Berkeley, California, University of California Press.

de Waal, F.B.M. [1998]: '"The social behavior of chimpanzees and bonobos: Empirical evidence and shifting assumptions": Reply', *Current Anthropology*, 39, pp. 407–408.

Diamond, J. [1991]: *The Rise and Fall of the Third Chimpanzee*, (1992) London, Vintage.

Dobzhansky, T.G. [1972]: 'Darwinian evolution and the problem of extraterrestrial life', *Perspectives in Biology and Medicine*, 15, pp. 157–176.

Dobzhansky, T.G. [1973]: 'Nothing in biology makes sense except in the light of evolution', *American Biology Teacher*, 35, pp. 125–129.

Double, R. [1990]: *The Non-reality of Free Will*, (1991) New York, Oxford University Press.

Double, R. [2002]: 'The moral hardness of libertarianism', *Philo*, 5, pp. 226–234.

Duhigg, C. [2019]: 'Is Amazon unstoppable?', *New Yorker*, 10 October, https://www.newyorker.com/magazine/2019/10/21/is-amazon-unstoppable

BIBLIOGRAPHY

Edelman, G.M. [1992]: *Bright Air, Brilliant Fire: On the Matter of the Mind*, New York, Basic Books.

Einstein, A. [1921]: 'Interview with the *Daily Mail* on the "Mystic Wireless"', *The Collected Papers of Albert Einstein, Volume 12: The Berlin Years: Correspondence January–December 1921*, https://einsteinpapers.press.princeton.edu/vol12-doc/508

Einstein, A. [1954]: *Ideas and Opinions*, trans. S. Bargmann, (1982) New York, Three Rivers Press.

Fitzgerald, M., Boddy, A. and Baum, S.D. [2020]: '2020 survey of artificial general intelligence projects for ethics, risk, and policy', *Global Catastrophic Risk Institute*, Technical Report 20-1, https://gcrinstitute.org/papers/055_agi-2020.pdf

F.L.I. [2016]: 'Benefits & risks of artificial intelligence', *Future of Life Institute*, June, https://futureoflife.org/background/benefits-risks-of-artificial-intelligence/

Fortey, R. [2000]: *Trilobite! Eyewitness to Evolution*, London, HarperCollins.

Fowler, A. and Hohmann, G. [2010]: 'Cannibalism in wild bonobos (Pan paniscus) at Lui Kotale', *American Journal of Primatology*, 72, pp. 509–514, DOI: 10.1002/ajp.20802

Fridman, L. [2021]: 'Avi Loeb: Aliens, black holes, and the mystery of the Oumuamua', *Lexfridman.com*, https://lexfridman.com/avi-loeb/ also https://www.youtube.com/watch?v=plcc6E-E1uU

Galton, F. [1857]: 'Negroes and the slave trade: To the Editor', *The Times*, 26 December, p. 10d, http://galton.org/essays/1850-1859/galton-1857-12-26-times-negroes-slave-trade.pdf

Galton, F. [1873]: 'Africa for the Chinese', Letter to *The Times*, 5 June, http://galton.org/letters/africa-for-chinese/AfricaForTheChinese.htm

Galton, F. [1892]: *Hereditary Genius: An Inquiry into its Laws and Consequences* (second edition), London, Macmillan and Co., http://galton.org/books/hereditary-genius/text/pdf/galton-1869-genius-v3.pdf. The almost identical 1869 first edition can be found here: http://galton.org/books/hereditary-genius/galton-1869-Hereditary_Genius.pdf.

Gould S.J. [1988]: 'Kropotkin was no crackpot', *Natural History*, 97, pp. 12–21.

Gould, S.J. [1993]: *Eight Little Piggies: Reflections in Natural History*, (1994) London, Penguin.

Goyal, A. [2003]: *Uncovering Russia*, Moscow, Norasco Publishing Limited.

Haldane, J.B.S. [1955]: 'Population genetics', *New Biology*, 18, pp. 34–51.

Hamai, M., Nishida, T., Takasaki, H. and Turner, L.A. [1992]: 'New records of within-group infanticide and cannibalism in wild chimpanzees', *Primates*, 33, pp. 151–162.

Hamilton, W.D. [1971]: 'Selection of selfish and altruistic behavior in some extreme models'. In J.F. Eisenberg and W.S. Dillon (*eds.*) *Man and Beast: Comparative Social Behavior*, Washington, Smithsonian Institution, pp. 57–92.

Hamilton, W.D. [1971a]: 'Geometry for the selfish herd', *Journal of Theoretical Biology*, 31, pp. 295–311.

Hamilton, W.D. [1975]: 'Innate social aptitudes of man: An approach from evolutionary genetics'. In W.D. Hamilton (*ed.*) *Narrow Roads of Gene Land: The Collected Papers of W.D. Hamilton, Volume I*, (1996) Oxford, W.H Freeman, pp. 329–351.

Hanson, R. [1998]: 'The Great Filter – Are we almost past it?', *George Mason University*, http://mason.gmu.edu/~rhanson/greatfilter.html

Harari, Y.N. [2018]: 21 Lessons for the 21st Century, London: Jonathan Cape.

Hawking, S., Russell, S., Tegmark, M. and Wilczek, F. [2014]: 'Stephen Hawking: "Transcendence looks at the implications of artificial intelligence – but are we taking AI seriously enough?"', *Independent*, 1 May, https://www.independent.co.uk/news/science/stephen-hawking-transcendence-looks-implications-artificial-intelligence-are-we-taking-ai-seriously-enough-9313474.html

Hiraiwa-Hasegawa, M. [1992]: 'Cannibalism among non-human primates'. In M.A. Elgar and B.J. Crespi (*eds.*) *Cannibalism: Ecology and Evolution Among Diverse Taxa*, Oxford, Oxford University Press, pp. 323–338.

Hobaiter, C., Schel, A.M., Langergraber, K. and Zuberbühler, K. [2014]: '"Adoption" by maternal siblings in wild chimpanzees', *PLoS One*, 9, e103777, DOI: 10.1371/journal.pone.0103777

Hölldobler, B. and Wilson, E.O. [1994]: *Journey to the Ants: A Story of Scientific*

Exploration, (1998) Cambridge, Massachusetts, Harvard University Press.

Horgan, J. [1995]: 'The new social Darwinists', *Scientific American*, 273, pp. 150–157.

Hrdy, S.B. [1977]: 'Infanticide as a primate reproductive strategy', *American Scientist*, 65, pp. 40–49.

Hrdy, S.B. [1977a]: *The Langurs of Abu: Female and Male Strategies of Reproduction*, (1980) Cambridge, Massachusetts, Harvard University Press.

Huxley, J. [1957]: 'Transhumanism'. In J. Huxley (*ed.*) *New Bottles for New Wine: Essays by Julian Huxley*, London, Chatto & Windus, pp. 13–17, https://archive.org/details/newbottlesfornew00juli

Huxley, T.H. [1894]: 'Evolution and ethics'. In J.G. Paradis and G.C. Williams (*eds.*) *Evolution & Ethics: T.H. Huxley's Evolution and Ethics*, (1989) Princeton, New Jersey, Princeton University Press, pp. 104–174.

I.S.S.N. [1920]: '"Hello, Earth! Hello!" Marconi believes he is receiving signals from the planets', *Idaho Springs Siftings-News*, 19 March, p.6, https://www.coloradohistoricnewspapers.org/?a=d&d=SSN19200319-01.2.58 and also see https://azmemory.azlibrary.gov/digital/collection/bensonsignal/id/1774

Izar, P., Verderane, M.P., Visalberghi, E., Ottoni, E.B., Gomes de Oliveira, M., Shirley, J. and Fragaszy, D. [2006]: 'Cross-genus adoption of a marmoset (*Callithrix jacchus*) by wild capuchin monkeys (*Cebus libidinosus*): Case report', *American Journal of Primatology*, 68, pp. 692–700.

Jacobs, J.A. [2001]: *Choosing Character: Responsibility for Virtue & Vice*, Ithaca, New York, Cornell University Press.

John, L.K., Loewenstein, G. and Prelec, D. [2012]: 'Measuring the prevalence of questionable research practices with incentives for truth telling', *Psychological Science*, 23, pp. 524–532; DOI: 10.1177/0956797611430953

Johnston, C. [2014]: 'Biological warfare flares up again between EO Wilson and Richard Dawkins', *Guardian*, 7 November, https://www.theguardian.com/science/2014/nov/07/richard-dawkins-labelled-journalist-by-eo-wilson

Jones, J.S. [1982]: 'Of cannibals and kin', *Nature*, 299, pp. 202–203.

Jones, J.S. [1999]: *Almost Like a Whale:* The Origin of Species *Updated*, London, Doubleday.

Jones, S., Winfield, A.F., Hauert, S. and Studley, M. [2019]: 'Onboard evolution of understandable swarm behaviors', *Advanced Intelligent Systems*, 1, 1900031, https://doi.org/10.1002/aisy.201900031

Judah, S. [2013]: 'Making time: Does it matter why we help others?', *BBC website*, 10 October, https://www.bbc.co.uk/news/magazine-24457645

Kahneman, D. [2011]: *Thinking, Fast and Slow*, London, Allen Lane.

Kano, T. [1998]: '"The social behavior of chimpanzees and bonobos: Empirical evidence and shifting assumptions": Reply', *Current Anthropology*, 39, pp. 410–411.

Kant, I. [1795]: 'Perpetual peace', trans. L.W. Beck. In Lewis White Beck (*ed.*) *Kant Selections*, (1988) London, Collier Macmillan, pp. 430–457.

Kelvin, Lord [1894]: *Popular Lectures and Addresses: Volume 2 Geology and General Physics*, (2011) Cambridge, Cambridge University Press.

Kropotkin, P. [1890]: 'Mutual aid among animals', *Nineteenth Century*, 28, pp. 337–354.

Kukuk, P.F. [1992]: 'Cannibalism in social bees'. In M.A. Elgar and B.J. Crespi (*eds.*) *Cannibalism: Ecology and Evolution Among Diverse Taxa*, Oxford, Oxford University Press, pp. 214–237.

Levy, N. [2011]: *Hard Luck: How Luck Undermines Free Will and Moral Responsibility*, Oxford, Oxford University Press.

Levy, N. [2019]: 'Taking responsibility for responsibility', *Public Health Ethics*, 12, pp. 103–113.

Lincoln, A. [1858]: 'Fourth debate with Stephen A. Douglas at Charleston, Illinois', Collected Works of Abraham Lincoln, Volume 3, *University of Michigan Digital Library Production Services*, 18 September, https://quod.lib.umich.edu/l/lincoln/lincoln3/1:20.1?rgn=div2;view=fulltext

Lloyd, S. [2012]: 'A Turing test for free will', *Philosophical Transactions of the Royal Society A*, 370, pp. 3597–3610, DOI: 10.1098/rsta.2011.0331

Loeb, A. [2021]: *Extraterrestrial: The First Sign of Intelligent Life Beyond Earth*, Boston, Mariner Books.

Lumsden, C.J. and Wilson, E.O. [1981]: *Genes, Mind, and Culture: The Coevolutionary Process*, Cambridge, Massachusetts, Harvard University Press.

Marin, J.M. [2009]: '"Mysticism" in quantum mechanics: The forgotten controversy', *European Journal of Physics*, 30, pp. 807–822.

Maynard Smith, J. [1964]: 'Group selection and kin selection', *Nature*, 201, pp. 1145–1147.

Maynard Smith, J. [1992]: *Did Darwin Get It Right? Essays on Games, Sex and Evolution*, London, Chapman & Hall.

Maynard Smith, J. [1993]: 'Confusion over evolution: An exchange', *New York Review of Books*, 14 January, p. 43.

Maynard Smith, J. [1996]: 'Conclusions'. In W.C. Runciman, J. Maynard Smith and R.I.M. Dunbar (*eds.*) *Evolution of Social Behaviour Patterns in Primates and Man: A Joint Discussion Meeting of the Royal Society and the British Academy*, and published as *Proceedings of the British Academy*, 88, pp. 291–297.

Maynard Smith, J. and Price, G.R. [1973]: 'The logic of animal conflict', *Nature*, 246, pp. 15–18.

Maynard Smith, J. and Szathmáry, E. [1995]: *The Major Transitions in Evolution*, Oxford, W.H. Freeman and Company.

Maynard Smith, J. and Szathmáry, E. [1999]: *The Origins of Life: From the Birth of Life to the Origins of Language*, Oxford, Oxford University Press.

Mayr, E.W. [1985]: 'The probability of extraterrestrial intelligent life'. In E.W. Mayr (*ed.*) *Toward a New Philosophy of Biology: Observations of An Evolutionist*, (1988) Cambridge, Massachusetts, Harvard University Press, pp. 67–74.

McCormick, R. [2016]: 'Elon Musk: There's only one AI company that worries me', *theverge.com*, 2 June, https://www.theverge.com/2016/6/2/11837566/elon-musk-one-ai-company-that-worries-me

Metz, R. [2022]: 'No, Google's AI is not sentient', *CNN Business website*, 14 June, https://edition.cnn.com/2022/06/13/tech/google-ai-not-sentient/index.html

Meyer, A. [2010]: 'George C. Williams (1926–2010)', *Nature*, 467, p. 790, https://www.nature.com/articles/467790a

Midgley, M. [1979]: 'Gene-juggling', *Philosophy*, 54, pp. 439–458.

Miles, J.B. [1998]: 'Unnatural selection', *Philosophy*, 73, pp. 593–608.

Miles, J.B. [2003]: *Born Cannibal: Evolution and the Paradox of Man*, London, IconoKlastic Books. Foreword by George C. Williams.

Miles, J.B. [2013]: '"Irresponsible and a disservice": The integrity of social psychology turns on the free will dilemma', *British Journal of Social Psychology*, 52, pp. 205–218.

Miles, J.B. [2015]: *The Free Will Delusion: How We Settled for the Illusion of Morality*, Kibworth Beauchamp, Matador.

Milton, K. [1998]: '"The social behavior of chimpanzees and bonobos: Empirical evidence and shifting assumptions": Reply', *Current Anthropology*, 39, pp. 411–412.

Mineau, P. and Cooke, F. [1979]: 'Rape in the lesser snow goose', *Behaviour*, 70, pp. 280–291.

Mitchell, M. [2021]: 'Why AI is harder than we think', *GECCO '21: Proceedings of the Genetic and Evolutionary Computation Conference*, June, https://arxiv.org/pdf/2104.12871.pdf

Mittal, D., Chakrabarti, S., Khambda, S.B. and Bump, J.K. [2020]: 'Spots and manes: The curious case of foster care between two competing felids', *Ecosphere*, 11, e03047. 10.1002/ecs2.3047

Mlodinow, L. [2012]: *Subliminal: How Your Unconscious Mind Rules Your Behavior*, (2013) New York, Vintage Books.

Musk, E. [2021]: 'Becoming multiplanetary…', Twitter.com, 22 May, https://twitter.com/elonmusk/status/1396226161718349824?lang=en

Neer, R.M. [2011]: 'Napalm, an American biography', Colombia University PhD dissertation, https://academiccommons.columbia.edu/doi/10.7916/D8R49Z3K/download. Subsequently published as the book of this title by Harvard University Press in 2013.

Nietzsche, F. [1886]: Beyond Good and Evil, trans. R.J. Hollingdale, (1990) London, Penguin.

Nowak, M.A., Tarnita, C.E. and Wilson, E.O. [2010]: 'The evolution of eusociality', Nature, 466, pp. 1057–1062, DOI:10.1038/nature09205

Nowak, M.A. and Highfield, R. [2011]: *SuperCooperators: Altruism, Evolution, and Why We Need Each Other to Succeed*, (2012) New York, Free Press.

Nuffield Council on Bioethics [2002]: *Genetics and Human Behaviour: The Ethical Context*, London, Nuffield Council on Bioethics.

OpenAI [2015]: 'Introducing OpenAI', *OpenAI.com*, 11 December, https://openai.com/blog/introducing-openai/

Paradis, J.G. and Williams, G.C. [1989]: *Evolution & Ethics: T.H. Huxley's 'Evolution and Ethics' With New Essays on its Victorian and Sociobiological Context*, Princeton, New Jersey, Princeton University Press.

Parfit, D. [2011]: *On What Matters, Volume One*, Oxford, Oxford University Press.

Pereboom, D. [2001]: *Living Without Free Will*, Cambridge, Cambridge University Press.

Pereboom, D. [2007]: 'Hard incompatibilism'. In J.M. Fischer, R. Kane, D. Pereboom and M. Vargas (*eds.*) *Four Views on Free Will*, Malden, Massachusetts, Blackwell, pp. 85–125.

Pinker, S. [1994]: *The Language Instinct: The New Science of Language and Mind*, (1995) London, Penguin.

Pinker, S. [1997]: *How the Mind Works*, (1998) London, Penguin.

Pinker, S. [1997a]: 'Evolutionary psychology: An exchange', *New York Review of Books*, 9 October, pp. 55–56.

Pinker, S. [2006]: 'Groups and genes: The lessons of the Ashkenazim', *New Republic*, 26 June, https://newrepublic.com/article/77727/groups-and-genes

Pinker, S. [2011]: *The Better Angels of Our Nature*, New York, Viking.

Pinker, S. [2012]: 'The false allure of group selection', *Edge.org*, 18 June, https://www.edge.org/conversation/steven_pinker-the-false-allure-of-group-selection

Pinker, S. [2015]: 'Thinking does not imply subjugating', *Edge.org*, https://www.edge.org/response-detail/26243

Piran, T. and Jimenez, R. [2014]: 'On the role of GRBs on life extinction in the universe', *Physical Review Letters*, 113, DOI: 10.1103/PhysRevLett.113.231102

Preston, P. [2012]: *The Spanish Holocaust: Inquisition and Extermination in Twentieth-Century Spain*, London, Harper Press.

Price, G.R. [1970]: 'Selection and covariance', *Nature*, 227, pp. 520–521.

Raby, P. [2001]: *Alfred Russel Wallace: A Life*, (2002) London, Pimlico.

Raulin-Cerceau, F. [2010]: 'The pioneers of interplanetary communication: From Gauss to Tesla', *Acta Astronautica*, 67, pp. 1391–1398. (This was a

Special Issue of *Acta Astronautica* on Searching for Life Signatures. Note: this paper sometimes referenced as Cerceau, F.)

Ridley, Mark and Dawkins, R. [1981]: 'The natural selection of altruism'. In J.P. Rushton and R.M. Sorrentino (*eds.*) *Altruism and Helping Behavior: Social, Personality and Developmental Perspectives*, Hillsdale, New Jersey, Lawrence Erlbaum, pp. 19–39.

Ridley, Matt [1996]: *The Origins of Virtue: Human Instincts and the Evolution of Cooperation*, Harmondsworth, Viking.

R.T. [2017]: '"Whoever leads in AI will rule the world": Putin to Russian children on Knowledge Day', *rt.com*, 1 September, https://www.rt.com/news/401731-ai-rule-world-putin/ (May need a VPN, as RT – formerly Russia Today – is directed by the Kremlin and so is now blocked in much of the West.)

Ruse, M. [1989]: *The Darwinian Paradigm: Essays on its History, Philosophy, and Religious Implications*, London, Routledge.

Russell, P. [1995]: *Freedom and Moral Sentiment: Hume's Way of Naturalizing Responsibility*, New York, Oxford University Press.

Sackur, S. [2016]: 'HARDtalk: Jim Ratcliffe, founder and chairman, Ineos', *BBC Radio 4*, 6 December, http://www.bbc.co.uk/programmes/b0854916

Sagan, C. [1980]: *Cosmos*, (1981) London, Book Club Associates.

Samuni, L., Wittig, R.M. and Crockford, C. [2019]: 'Adoption in the Taï chimpanzees: Costs, benefits, and strong social relationships'. In Boesch and others (*eds.*) *The Chimpanzees of the Taï Forest: 40 Years of Research*, Cambridge, Cambridge University Press, pp. 141–158.

Sander, P.M. [2013]: 'An evolutionary cascade model for sauropod dinosaur gigantism – Overview, update and tests', *PLoS One*, 8, e78573, DOI:10.1371/journal.pone.0078573

Sander, P.M., Christian, A., Clauss, M., Fechner, R., Gee, C.T., Griebeler, E., Gunga, H., Hummel, J., Mallison, H., Perry, S.F., Preuschoft, H., Rauhut, O.W.M., Remes, K., Tütken, T., Wings, O. and Witzel, U. [2011]: 'Biology of the sauropod dinosaurs: The evolution of gigantism', *Biological Reviews*, 86, pp. 117–155.

Searle, J.R. [2000]: 'Consciousness, free action and the brain', *Journal of Consciousness Studies*, 7, pp. 3–22.

Segerstrale, U. [2000]: *Defenders of the Truth: The Sociobiology Debate*, Oxford, Oxford University Press.

Sherwood, C.C., Miller, S.B., Karl, M., Stimpson, C.D., Phillips, K.A., Jacobs, B., Hof, P.R., Raghanti, M.A., and Smaers, J.B. [2020]: 'Invariant synapse density and neuronal connectivity scaling in primate neocortical evolution', *Cerebral Cortex*, 30, pp. 5604–5615.

Sibley, C.G. and Ahlquist, J.E. [1984]: 'The phylogeny of the hominoid primates, as indicated by DNA-DNA hybridization', *Journal of Molecular Evolution*, 20, pp. 2–15.

Smilansky, S. [2000]: *Free Will and Illusion*, Oxford, Oxford University Press.

Smilansky, S. [2011]: 'Free will, fundamental dualism, and the centrality of illusion'. In R. Kane (*ed.*) *The Oxford Handbook of Free Will, Second Edition*, Oxford, Oxford University Press, pp. 425–441.

Sober, E. and Wilson, D.S. [1998]: *Unto Others: The Evolution and Psychology of Unselfish Behavior*, (1999) Cambridge, Massachusetts, Harvard University Press.

Sommer, V. [2000]: 'The holy wars about infanticide. Which side are you on? And why?'. In C.P. van Schaik and C.H. Janson (*eds.*) *Infanticide by Males and its Implications*, Cambridge, Cambridge University Press, pp. 9–26.

Stanford, C.B. [1998]: 'The social behavior of chimpanzees and bonobos: Empirical evidence and shifting assumptions', *Current Anthropology*, 39, pp. 399–420.

Strawson, G. [1998] 'Luck swallows everything: Can our sense of free will be true?', *Times Literary Supplement*, 26 June, pp. 8–10.

StuChannel3 [2010]: 'Robert Winston – The Jim Twins', *Youtube.com*, 27 December, https://www.youtube.com/watch?v=qw3S35wGgT8

Sullivan, A. [2011]: 'The study of intelligence', *Andrewsullivan.com*, 21 November, http://dish.andrewsullivan.com/2011/11/21/the-study-of-intelligence/

Sullivan, A. [2011a]: 'The study of intelligence (continued)', *Andrewsullivan.com*, 28 November, http://dish.andrewsullivan.com/2011/11/28/the-study-of-intelligence-ctd-1/

Symons, D. [1992]: 'On the use and misuse of Darwinism in the study of human behavior'. In Barkow and others (*eds.*) *The Adapted Mind: Evolutionary*

Psychology and the Generation of Culture, New York, Oxford University Press, pp. 137–159.

Szathmáry, E. and Maynard Smith, J. [1995]: 'The major evolutionary transitions', *Nature*, 374, pp. 227–232.

Taylor, C. [1985]: *Philosophical Papers: Volume 1, Human Agency and Language*, (1999) Cambridge, Cambridge University Press.

Tokuyama, N., Moore, D.L., Graham, K.E., Lokasola, A. and Furuichi, T. [2017]: 'Cases of maternal cannibalism in wild bonobos (Pan paniscus) from two different field sites, Wamba and Kokolopori, Democratic Republic of the Congo', *Primates*, 58, pp. 7–12, https://doi.org/10.1007/s10329-016-0582-7

Tokuyama, N., Toda, K., Poiret, M.L., Iyokango, B., Bakaa, B. and Ishizuka, S. [2021]: 'Two wild female bonobos adopted infants from a different social group at Wamba', *Scientific Reports*, 11, 4967, https://doi.org/10.1038/s41598-021-83667-2

Trivers, R.L. [1971]: 'The evolution of reciprocal altruism', *Quarterly Review of Biology*, 46, pp. 35–57.

Trivers, R.L. [1985]: *Social Evolution*, Menlo Park, California, Benjamin/Cummings Publishing.

Tuci, E., Ampatzis, C., Vicentini, F. and Dorigo, M. [2008]: 'Evolving homogeneous neurocontrollers for a group of heterogeneous robots: Coordinated motion, cooperation, and acoustic communication', *Artificial Life*, 14, pp. 157–178.

Tung, L. [2016]: 'Google Alphabet's Schmidt: Ignore Elon Musk's AI fears – he's no computer scientist', *zdnet.com*, 9 June, https://www.zdnet.com/article/google-alphabets-schmidt-ignore-elon-musks-ai-fears-hes-no-computer-scientist/

Turing, A.M. [1951]: 'Intelligent machinery, a heretical theory'. In S.M. Shieber (*ed.*) *The Turing Test: Verbal Behavior as the Hallmark of Intelligence*, (2004) Cambridge, MIT Press, pp. 105–109.

Tyrrell, T. [2020]: 'Chance played a role in determining whether Earth stayed habitable', *Communications Earth & Environment*, 1, pp. 1-10, https://doi.org/10.1038/s43247-020-00057-8

U.J.A. [2018]: 'Together and apart: Israeli Jews' views on their relationship to American Jews and religious pluralism', *Ujafedny.org*, 29 January, https://www.ujafedny.org/api/v2/assets/789241/

Uriarte, A., Johnstone, C., Laiz-Carrión, R., García, A., Llopiz, J.K., Shiroza, A., Quintanilla, J.M., Lozano-Peral, D., Reglero, P. and Alemany, F. [2019]: 'Evidence of density-dependent cannibalism in the diet of wild Atlantic bluefin tuna larvae (*Thunnus thynnus*) of the Balearic Sea (NW-Mediterranean)', *Fisheries Research*, 212, pp. 63–71.

van Noordwijk, M.A. and van Schaik, C.P. [2000]: 'Reproductive patterns in eutherian mammals: Adaptations against infanticide?' In C.P. van Schaik and C.H. Janson (*eds.*) *Infanticide by Males and its Implications*, Cambridge, Cambridge University Press, pp. 322–360.

Varki, A. and Nelson, D.L. [2007]: 'Genomic comparisons of humans and chimpanzees', *Annual Review of Anthropology*, 36, pp.191–209.

vonHoldt, B.M., Shuldiner, E., Janowitz Koch, I. *et al.* [2017]: 'Structural variants in genes associated with human Williams-Beuren syndrome underlie stereotypical hypersociability in domestic dogs', Science Advances, 3, e1700398, DOI: 10.1126/sciadv.1700398

Wade, N. [2010]: 'George C. Williams, 83, theorist on evolution, dies', *New York Times*, 13 September, https://www.nytimes.com/2010/09/14/science/14williams.html

Wakefield, M.L., Hickmott, A.J., Brand, C.M., Takaoka, I.Y., Meador, L.M., Waller, M.T. and White, F.J. [2019]: 'New observations of meat eating and sharing in wild bonobos (*Pan paniscus*) at Iyema, Lomako Forest Reserve, Democratic Republic of the Congo', *Folia Primatologica*, 90, pp. 179–189.

Walker, M. [2010]: 'Wild bonobo mother ape eats own infant in DR Congo', *BBC website*, 1 February, http://news.bbc.co.uk/earth/hi/earth_news/newsid_8487000/8487138.stm

Wallace, A.R. [1891]: *Natural Selection and Tropical Nature: Essays on Descriptive and Theoretical Biology*, (1969) Farnborough, Gregg International.

Waller, B.N. [1990]: *Freedom Without Responsibility*, Philadelphia, Temple University Press.

Waller, B.N. [2006]: 'Denying responsibility without making excuses', *American Philosophical Quarterly*, 43, pp. 81–90.

Waller, B.N. [2011]: *Against Moral Responsibility*, Cambridge, Massachusetts, MIT Press.

Waller, B.N. [2012]: 'Exchange on Bruce Waller's *Against Moral Responsibility*'. Available at https://dl.tufts.edu/pdfviewer/08613068g/9w032f65c

Waller, B.N. [2015]: *The Stubborn System of Moral Responsibility*, Cambridge, Massachusetts, MIT Press.

Warner, B. [2009]: 'Charles Darwin and John Herschel', *South African Journal of Science*, 105, pp. 432–439, http://www.scielo.org.za/pdf/sajs/v105n11-12/a1405112.pdf

Watson, G. [2004]: 'Responsibility and the limits of evil: Variations on a Strawsonian theme'. In G. Watson (*ed.*) *Agency and Answerability: Selected Essays*, New York, Oxford University Press, pp. 219–259.

Wilde, O. [1905]: *De Profundis*, (1996) New York, Dover Publications Inc.

Williams, G.C. [1966]: *Adaptation and Natural Selection: A Critique of Some Current Evolutionary Thought*, (1996) Princeton, New Jersey, Princeton University Press.

Williams, G.C. [1988]: 'Huxley's "Evolution and Ethics" in sociobiological perspective', *Zygon*, 23, pp. 383–407.

Williams, G.C. [1996]: *Plan & Purpose in Nature*, (1997) London, Phoenix.

Wilson, D.S. [2015]: 'The tide of opinion on group selection has turned', *Evolution Institute*, 26 May, https://evolution-institute.org/blog/the-tide-of-opinion-on-group-selection-has-turned/

Wilson, E.O. [1975]: *Sociobiology: The New Synthesis*, Cambridge, Massachusetts, Harvard University Press.

Wilson, E.O. [1975a]: 'Human decency is animal', *New York Times Magazine*, 12 October, pp. 38–50, https://www.nytimes.com/1975/10/12/archives/human-decency-is-animal-hawks-and-baboons-are-not-usually-heroic.html

Wilson, E.O. [1978]: *On Human Nature*, Cambridge, Massachusetts, Harvard University Press.

Wright, L. [1997]: *Twins: Genes, Environment and the Mystery of Identity*, London, Weidenfeld & Nicolson.

Wright, L. [1997a]: 'Twins prove life's a script', *The Times*, 3 November, p. 15.

Wright, R. [1994]: *The Moral Animal: Evolutionary Psychology and Everyday Life*, (1996) London, Abacus.

Wright, R. [1999]: 'The accidental creationist: Why Stephen Jay Gould is bad for evolution', *New Yorker*, 13 December, pp. 56–65. See https://robertwright.com/accidental-creationist/

Wynne-Edwards, V.C. [1962]: *Animal Dispersion in Relation to Social Behaviour*, Edinburgh, Oliver and Boyd.

Yasseri, T. [2016]: 'P-values are widely used in the social sciences, but often misunderstood: And that's a problem', *Oxford Internet Institute*, 7 March, http://blogs.oii.ox.ac.uk/policy/many-of-us-scientists-dont-understand-p-values-and-thats-a-problem/

Yong, E. [2012]: 'Nobel laureate challenges psychologists to clean up their act', *Nature*, 3 October, https://www.nature.com/articles/nature.2012.11535

Zwaan, R.A. [2013]: 'Beware of Voodoo experimentation', 21 March, https://rolfzwaan.blogspot.co.uk/2013/03/beware-of-voodoo-experimentation.html

Zyga, L. [2009]: 'Quantum mysticism: Gone but not forgotten', *Phys.org*, 8 June, https://phys.org/news/2009-06-quantum-mysticism-forgotten.html

INDEX

Adaptation, 4, 7, 24, 31–2, 35–6, 38–9, 43–5, 47–50, 59, 72, 102, 106, 152, 251
Adaptation and Natural Selection, 4, 35, 37–8, 48, 252
Adelphophagy, 61
Adoption, intergenus, 68
AI (artificial intelligence)
    AGI, 145–58, 168–71
    baked-in problem of, 155, 158, 171
    existential threat of, 4, 18, 24, 141–160, 170–1, 228–9, 233
    human-level AI as an oxymoron, 158
Alexander, Richard, 46, 235
Alien, 92, 192, 224
Altruism
    kin selection, 44, 50, 54–6, 73, 76, 109, 156, 161, 196
    reciprocal altruism, 37, 45–7, 49, 54–6, 71, 73–4, 76, 96, 151, 156–7, 189, 201
American Statistical Association, 202–3, 235
Analysis versus "truer" analysis, 131–2, 184, 191
Ant, driver, 80–1, 119, 171
Archosaur, crocodile-line, 23, 99, 163
Asimov, Isaac, 150
Auerbach, David, 168–70, 235
Australopithecus, 105
"A world without compassion" (chimpanzee), 36–7, 42, 47–8, 70, 99–100, 103, 105, 109

Baboon, 35, 74, 119
Barbarous Years, The, 179
Barkow, Jerome, 48, 97, 236
Behavioural genetics, two-stage developmental process, 204–9

Bennett, Nigel C., 77, 79, 236
Better Angels of Our Nature, The, 121, 247
Bezos, Jeff, 185–7, 232, 237
Blasdel, Alex, 130–1, 237
Bonobo
    adoption, 68–9, 215–6
    cannibalism, 67–8, 82, 100, 171, 216
Bostrom, Nick, 147–8, 155, 161–2, 166, 169–70, 237
Bouchard, Thomas J., 209–11
Britten, Roy, 104–5
Brown, Andrew, 196–7, 201, 237
Burgess Shale, 199
Bygott, J. David, 57–9, 187, 238

Cameron, James, 176
Cannibalism, 49, 57–9, 61–2, 64, 67–71, 78–9, 82–3, 85–6, 103, 105–6, 174, 180, 187, 190, 206–7, 216
Causa sui, 111
Chimpanzee, 33, 36–7, 42, 47, 57–9, 62–6, 69–71, 73–5, 82, 84–6, 95, 99–108, 119, 158, 173–4, 189, 215–6
Confabulation, 129
Convergent evolution, 20–1, 220
Conway Morris, Simon, 199, 238
Cornell Evolution Project, a Survey of Evolution and Religious Belief, 126, 130, 132
Cosmic outsider, 110, 113, 116, 136, 181, 218, 225
Cosmides, Leda, 48, 97, 196, 238
Cotterill, Sarah, 177, 238
Cox, Brian, 1, 3, 12–3, 16, 53, 75, 83, 221, 238
Crafoord Prize, 5, 10, 198, 219
Crespi, Bernard J., 78–80, 238
Crick, Francis, 88, 130
Cronin, Helena, 24, 29, 34, 49, 85, 90, 98, 106, 109, 239

Dalit, out-of-caste, 176–7
Darwin, Charles, 2–6, 8–18, 24–38, 45, 49, 51, 54–7, 62, 71, 75–6, 82–4, 86–92, 94–7, 103, 107–8, 111, 115–6, 120–1, 129, 146, 150–2, 156, 160, 173, 180, 182, 188, 190–2, 194–6, 198, 200–1, 203–4, 219, 222–3, 232–4, 239
Darwin, Erasmus, 30
Darwin's bulldog. *See* Huxley, Thomas H.
Darwin wars, the, 195–9
Dawkins, Richard, 21, 28, 36, 41, 43, 50–1, 53, 62, 64, 80–1, 85, 88, 99–100, 102, 107, 118–9, 124, 130–1, 156, 171, 184, 188, 194–6, 198–9, 201, 230, 239
Dellatore, David, 67, 239
Dennett, Daniel C., 95, 131, 133–8, 183, 185, 194, 201, 204, 239–40
Descent of Man, The, 35, 55, 62, 92, 108, 188, 239
Determinism, dilemma of, 126–7, 138
de Waal, Frans, 36, 42, 64–6, 69–70, 99, 101, 108, 146, 240
Diamond, Jared, 104–6, 240
Dinosaur
    cosmic, 1, 14, 17–8, 24–5, 27, 29, 53, 74, 83, 89, 91, 116, 150, 163, 191, 215, 222, 232–3

INDEX

gigantism of, 21–3, 163
Dobzhansky, Theodosius, 9, 209–11, 219–22, 240
Dog, hypersociability of, 63, 207
Domestication
　of dog, 63–4, 68, 153
　of human, 207–8
Double dose of unfairness animal, the, 182–7, 204
Double, Richard, 127, 240
Doubly unfair ape, the, 182–7
Dragonfly, extinct giant (Meganisoptera), 22, 191

Eclipse of Darwinism, the, 32
Edelman, Gerald, 132, 241
Edison, Thomas, 2, 225
Einstein, Albert, 2, 129, 241
Eldredge, Niles, 195–6
Epigenesis, 8, 103
Essentialism, philosophical, 137, 183
E.T.
　homicidal, 75, 82–8, 92, 94,
　Type I, 53–94, 110, 115–7, 119–20, 150, 167, 173, 192, 228, 231, 233
　Type II, 83–4, 94–117, 120, 126, 133, 136, 138–40, 150, 164, 167, 172–3, 187–8, 192, 200, 213–8, 225–9, 231, 233–4
Eugenics, paradox of, 207
Eusociality, 6–7, 12, 28, 50–1, 55, 77–81, 84, 94, 109, 115, 215, 223
Evolutionarily stable strategy (ESS), 50, 72, 118, 173, 189
Evolutionary psychology (EP), 8, 47–50, 98, 199–200

Evolutionary swarm robotics, 86, 89, 158–60
Exaggerated gregariousness. *See* Dog, hypersociability of
Exceptionalism, 11

Fatalism, contrasted with determinism, 111–2
Faulkes, Chris G., 77, 79, 236
Fermi paradox, 1, 19, 222, 224
Fieser, Louis, 87
Fisher, Ronald A., 35
Fortey, Richard, 199, 241
Fossey, Dian, 58, 71
Fowler, Andrew, 67, 71, 241
Free will, 126–39, 183
Fridman, Lex, 115, 162, 230, 241

Galton, Sir Francis, 8–9, 181–2, 190–1, 206–7, 209, 217, 241
Game theory, 5, 10, 71–2, 119, 124, 169
Gates, Bill, 121, 131, 145, 175, 181
Gauss, Johann Carl Friedrich, 1–2
Gene, definition within evolutionary theory, 36
Ghost elbow, 135
Gigerenzer, Gerd, 126
Gladstone, John, 176
Global Catastrophic Risk Institute, 145–7
Goodall, Jane, 58
Gorilla, 33, 35, 58, 66, 71, 104
Gould, Stephen Jay, 37–8, 40, 43, 49–50, 59, 195–7, 199, 241–2
Graffin, Greg, 130
Great Filter, the, 2, 18–9
Great Silence, the, 1–2, 5, 13, 17–8,

87, 89, 91, 215, 224, 226–8
Gull, blackheaded, 62, 118

Haldane, J.B.S., 35, 55, 161, 242
Hamilton, William D., 10, 36–7, 41–2, 45, 49–50, 54–5, 60, 73, 76, 108–9, 118, 153, 161, 172, 194–8, 242
Hanson, Robin, 18–9, 242
Haplodiploidy sex determination system, 55, 76, 156, 189
Harari, Yuval Noah, 127, 242
Harvard University, tragedy of, 88, 156
Hawking, Stephen, 141, 145–8, 155, 242
Herschel, Sir John, 34, 191
Hide (bonobo maternal cannibalism), 67–8, 82, 92, 151, 167, 171, 189, 224
Higgs boson, 203
"Highest & most interesting problem for the naturalist" (Darwin), 24–5, 27–8, 200, 230, 232
Hippocampus question, the, 33
Hiraiwa-Hasegawa, Mariko, 58, 101, 242
Hohmann, Gottfried, 67, 71, 241
Hölldobler, Berthold K., 80–2, 84–5, 160, 171, 218, 230, 242
Homo erectus, 105
Homo habilis, 105
Homo sapiens, anatomically modern
    as best and smartest that can evolve, 4, 120, 122–6, 140, 222, 234
    limited rationality crucial to civilisation, 136, 139–40, 156–7

Horgan, John, 48, 98, 200, 243
Hrdy, Sarah B., 57–8, 101, 243
Human sociobiology, 6, 8–17, 24–5, 28–9, 34, 42–53, 69, 75, 81–6, 89–92, 95–103, 106–10, 116, 120–1, 130–1, 146, 150–3, 172–4, 180, 182, 184, 187–8, 194, 196–201, 203, 209, 216, 224, 228, 233–4
Humphrey (wild chimpanzee), 57, 82, 84–5, 92, 95, 98, 151, 171, 187–9, 215–7, 234
Huxley, Julian, 160–1, 243
Huxley, Thomas H., 2, 23, 33, 37–8, 47, 94–5, 108, 160, 172, 243
Hymenoptera, 76

Imperial Japanese army, atrocities of, 178
Inbreeding and outbreeding, 72–3, 76–8, 119, 156, 189, 223
Inclusive fitness, 10, 48, 50, 54–5, 76, 161
Indeterminism, dilemma of, 126
Indian caste system, 176–7, 184, 191
Infanticide, as adaptive behaviour, 57–8, 60, 65–7, 69, 103, 105
Inverse problem, the (Wallace), 25, 27, 200, 232
IQ, natural, 174–5, 205
Isoptera, 76
Israel, State of, 179

Jacobs, Jonathan, 135, 243
Jim twins, the, 209–11
Jones, Steve, 59, 63–4, 69, 154, 207, 243

INDEX

Kahneman, Daniel, 125–6, 128, 201–2, 212, 244
Kano, Takayoshi, 66, 101, 244
Kant, Immanuel, 111, 118–20, 122, 165, 244
Kardashev scale of galactic civilisation, 114–5, 225
Kelvin, Lord, 6, 8–9, 11, 32–4, 46–7, 51, 95, 121, 188, 244
Kin selection. *See* Altruism and
Kropotkin, Peter (Pyotr), 37–8, 43, 49, 51, 55, 90, 95, 99, 121, 195, 244
Kukuk, Penelope, 78–9, 244

Lamarckism, 30
Langur monkeys, 57, 101
LessWrong, 168–70
Levy, Neil, 182–5, 244
Lewontin, Richard, 40–1, 196–8
Libet, Benjamin, 129
Lincoln, Abraham
    as a white supremacist, 113, 121, 181, 190–1, 217, 244
    support for anti-miscegenation laws, 113, 190
    the Great Emancipator, 113, 181
Loeb, Avi, 19, 114–7, 149, 162–3, 225, 230, 237, 244
"Luck swallows everything", 127–8, 134, 137–9, 184
Lumsden, Charles, 44, 46, 244

Major Transitions in Evolution, The, 5–8, 12–4, 25, 88, 115, 220–3, 245, 250
Mallard ducks, rape within, 61

Marconi, Guglielmo, 2
Margulis, Lynn, 6, 12, 130
Marin, Juan, 132, 244
Mass extinctions, 23, 163, 220
Maynard Smith, John, 5–8, 10, 13, 16, 18, 36–41, 43–6, 48, 51, 54–6, 71–2, 79, 88, 99, 102, 107–8, 118–9, 122, 124, 130, 161, 169, 172–3, 194–6, 198–200, 206, 217, 219–22, 245
Mayr, Ernst, 198, 219, 221–2, 245
Megalodon (*Otodus megalodon*), 21–2, 163
Mendel, Gregor, 9, 32, 198, 219
Midgley, Mary, 201, 245
Milton, Katharine, 65–6, 246
Mitchell, Melanie, 144, 159, 246
Modern synthesis, the, 9, 32, 198, 219
Moffitt, Terrie, 213
Mole-rat
    Damaraland, 77, 223
    naked, 77–9, 81–2, 84, 86, 118–9, 223
Moral Animal, The, 47, 98, 182, 253
Moral luck, problem of, 127–8, 137–9, 183–4
Morgan, Thomas Hunt, 32
Mozu (deformed snow monkey), 42, 184, 192
Multilevel selection (MLS), 50, 109, 196
Musk, Elon, 18–9, 145–8, 154–5, 157, 163, 170, 230, 246

Napalm, 87–8, 90, 109, 156, 160, 229, 246
Natura non facit saltum, 102–3

New Optimists, the, 131
Nietzsche, Friedrich, 111, 129, 166, 246
Nowak, Martin, 50, 109, 193, 246
Nuffield Council on Bioethics, 212–3, 246

Orang-utan, 33, 35, 66–7, 69, 71
Origin of Species, On the, 30, 55, 63–4, 103, 239
Origins of Life, The, 6–8, 12–4, 25, 88, 115, 220–3, 245
'Oumuamua (space rock 1I/2017 U1), 26, 114, 225, 232
Owen, Sir Richard, 8–9, 11, 33–4, 188

Paley, William, 31–2
Pan paniscus. *See* Bonobo
Pan troglodytes. *See* Chimpanzee
Paradis, James, 59–60, 247
Parasitoids, koinobiont, 92
Parfit, Derek, 111–2, 247
Parish, Amy, 70, 146
Parsimony, principle of, 12, 15–7, 38, 51, 95–8, 152, 190, 221
Parthenogenesis, 7, 77, 156
Pauli, Wolfgang, 132
Pereboom, Derk, 127, 247
Phylogenetic constraint, 19–20
Pinker, Steven, 9, 48–50, 91, 109, 120–1, 130–1, 134, 145–6, 174–5, 180, 185, 187–8, 203, 205–6, 210, 212, 247
Posthumanism. *See* Transhumanism
Pratchett, Terry, 24
Price equation, the, 10
Price, George R., 10, 36–7, 71, 108, 121, 167, 205, 245, 247
Prisoner's dilemma, iterated, 71, 124
Provine, William B., 126, 130, 134
Punctuated equilibrium, 195
P-values, statistical misunderstanding of, 202–3, 210

Quantum mysticism, 132

Race of "better angels" (sociobiology), 120–1, 172–4
Race of devils (*Volk von Teufeln*) problem, 117–22, 155–6, 165–6, 173
Race science, 9, 131, 174–5, 180–1, 190, 203–7
Rape, of female snow goose, 60–1
Ratcliffe, Sir Jim, 185–6, 232
Rationality, bounded, 126
Reason, 117–40, 145–9, 153–4, 157, 164–5, 172, 174, 181, 183, 185
Reciprocal altruism. *See* Altruism and
Reproducibility Project, the, Center for Open Science-led, 200, 202, 213
Reproductive skew, 79–80
Ridley, Mark, 41, 100, 248
Ridley, Matt, 15, 248
Roko's Basilisk, 168–71
Rose, Steven, 198
Ruse, Michael, 44–7, 248

Sagan, Carl, 18, 117, 172, 219, 225, 227, 248
Saltationism, 103, 107, 189
Sanchita karma, 177, 184
Schmidt, Eric, 146–7

Science, hard versus soft, 9, 200, 210
Scientific racism. *See* Race science
Seal, elephant, 62, 101
Searle, John, 138, 248
Segerstrale, Ullica, 43, 46, 198–9, 249
Segregation, in Northern US, 180
Selection, levels of
    genic (selfish gene-ery), 4–6, 8, 10, 17, 24–5, 28, 35–7, 40, 48, 51, 54–62, 65–74, 80, 83, 89, 95, 118, 150, 233
    group, 15–7, 24–5, 30, 36–42, 44–5, 47–8, 50–2, 64, 72–3, 83–4, 89, 95, 97, 99, 109, 116, 150, 193, 196–8, 233–4
    individual, 4, 10, 30, 33, 36, 40, 48, 54, 57, 62–3, 69–70, 80, 84, 112, 189
Selfish gene theory. *See* Selection, levels of, genic
Selfish herd theory, 73, 118
Shark, sand tiger, 61–2
Simon, Herbert A., 126
Singularity, technological, 147–9, 154, 168, 171, 228
Smilansky, Saul, 133, 138, 204, 249
Sober, Elliott, 15, 108–9, 151, 249
Social psychology, 202, 212–3
Sociobiology. *See* Human sociobiology
Sommer, Volker, 65, 249
Spanish Holocaust, The, 178–9
Spielberg, Steven, 92, 224
Stanford, Craig, 65, 249
Statistical relevance. *See P*-values
Strawson, Galen, 127, 184, 249
Subversion from within, 36, 41, 45, 72, 74, 79–81

Sullivan, Andrew, 174, 203, 205, 249
Swarm robotics. *See* Evolutionary swarm robotics
Symons, Donald, 48–9, 85, 91, 152, 249
Szathmáry, Eörs, 5–7, 13, 18, 79, 88, 220, 245, 250

Taylor, Charles, 135, 250
Tesla, Nikola, 2
Thiel, Peter, 169
Thomson, William. *See* Kelvin, Lord
Tokuyama, Nahoko, 67–9, 250
Tooby, John, 48, 97, 196, 236, 238
Transhumanism, 19, 141–2, 160–70, 226, 234
Transhumanist Declaration, 162
Transition, major evolutionary
    Darwin's eighth, 1, 6, 8, 12–4, 16, 26–9, 82, 84, 89, 91, 95–6, 141, 154, 208, 221–3, 228–30
    one through six, 6–8, 12–8, 25, 29, 75, 82, 84, 89, 94, 173, 192, 215, 221–3, 232
    seven (eusocial colonies), 6–7, 15, 25, 29, 75–82, 84, 86, 89, 92, 94, 119, 173, 215, 223, 232
Trivers, Robert L., 9, 28, 43, 45–7, 85, 121, 188, 250
Tuna, Atlantic bluefin (*Thunnus thynnus*), 62
Turing, Alan, 193, 202, 229, 250

Wallace, Alfred Russel, 24–5, 30–2, 34, 37, 54, 92, 200, 232, 251

Waller, Bruce, 127, 134–6, 138, 212, 251–2
Watson, Gary, 135, 137–8, 183, 252
Wells, Herbert G., 117, 225
Wilde, Oscar, 111, 136, 252
Williams, George C., 4–5, 7, 10, 16, 19–20, 35–41, 45, 48, 51, 54, 56–7, 59–61, 68, 70–2, 75, 81, 84, 100–1, 107, 111, 122, 127, 130, 147, 156, 185, 188–9, 194–6, 198, 217, 219, 252
Wilson, David Sloan, 15, 28, 40, 50, 108–9, 151, 249, 252
Wilson, Edward O., 10–1, 16, 28, 42–7, 50–1, 81, 98, 109, 121, 130–1, 182, 197–8, 252

Winston, Lord Robert, 210–1
World Transhumanist Association, 161–2
Wright, Lawrence, 209–10, 252–3
Wright, Robert, 47–9, 98, 182, 197, 253
Wright, Sewall, 35
Wynne-Edwards, Vero C., 38–9, 118, 195, 253

Yasseri, Taha, 202, 253
Yudkowsky, Eliezer, 169–70

Zuckerberg, Mark, 121, 131
Zwangsarbeit, slave labour, 177–8